AGRO-WASTE ALCHEMY

Transforming Agricultural residues into Sustainable value

Ms. Sreebavya V. S. | Mr. Sukumar S.

Mr. Shaju M. J. | Mrs. P. Vishnupriya

Ms. Dhanusha R. | Ms. Chandana C. | Mr. Lokanesh A.

Made with ♥ on the Notion Press Platform

www.notionpress.com

Dedication

To Our Parents

With profound gratitude and heartfelt respect, we dedicate this book to our parents, whose unwavering love, sacrifices, guidance, and encouragement have been the foundation of our personal and professional journeys. Their values, perseverance, and belief in education have inspired us to pursue knowledge, embrace innovation, and contribute meaningfully to society.

To Our Readers

We also dedicate this book to the students, researchers, academicians, industry professionals, and innovators who strive to create a more sustainable future. Through our academic pursuits, research endeavours, industrial experiences, and interactions with communities, we have witnessed both the challenges posed by agricultural waste and the immense opportunities hidden within it. These experiences have shaped our vision and motivated us to explore sustainable pathways for transforming agricultural residues into valuable resources. This book is a reflection of that journey and our commitment to advancing the principles of circular economy and resource sustainability.

To Future Generations

We dedicate this work to the generations yet to come. May they inherit a world where innovation and environmental stewardship go hand in hand, where waste is viewed not as a burden but as a resource, and where sustainable solutions drive economic growth and ecological balance. We hope this book inspires future thinkers, researchers, and leaders to continue developing transformative technologies that create a cleaner, greener, and more resilient planet.

May the knowledge share in these pages contribute to a sustainable future and encourage collective efforts toward responsible resource utilization and environmental conservation.

With sincere appreciation and gratitude,

The Authors

Agro-Waste Alchemy

Transforming Agricultural Residues into Value

ABOUT THE AUTHORS

First author

Sreebavya V. S. is an Assistant Professor and Research & Development Faculty in the Department of Chemical Engineering at Erode Sengunthar Engineering College, Tamil Nadu, India, and a Ph.D. scholar at National Institute of Technology Andhra Pradesh. She holds an M.A. in Environmental Studies from Tel Aviv University, an M.Tech. in Computer Aided Process Design from Government Engineering College Kozhikode, and a B.Tech. in Chemical Engineering from Government Engineering College Thrissur. Her research interests include sustainable process engineering, agro-waste valorization, bioplastics, environmental technologies, and circular bioeconomy innovations. She is committed to developing sustainable solutions that bridge scientific research and industrial applications.

Second Author

Mr. Sukumar S. is a Production Engineer at Elgi Rubber Company Ltd., India. He holds a Bachelor's degree in Chemical Engineering and has professional experience in industrial manufacturing and process operations. His areas of interest include process optimization, sustainable production practices, and industrial engineering applications. He is committed to advancing innovative and efficient technologies for sustainable industrial development.

Third Author

Shaju M.J. is from Kanniyakumari, Tamil Nadu, and holds a B.Tech in Chemical Engineering from Karpagam Academy of Higher Education, Coimbatore, and an M.S. in Chemical Engineering from Ming Chi University of Technology, Taiwan. His research focuses on sustainable energy, catalyst development, and metal recovery from spent lithium-ion batteries.He works on green catalytic applications and circular economy approaches.

He received an Honorable Mention Award at the International Conference SDSE 2025 for his research contribution.

Fourth Author

Mrs. Vishnupriya is an Assistant Professor in the Department of Biotechnology with around three years of teaching experience. Her academic interests include biotechnology, sustainability, and waste valorisation. She is passionate about promoting innovative solutions for environmental challenges through education and research.

Fifth Author:

Dhanusha R is an Assistant Professor in the Department of Food Technology at Dhanalakshmi Srinivasan College of Engineering. She holds a B.Tech. in Food Technology and an M.Tech. in Food Process Engineering and is currently pursuing a part-time Ph.D. in Chemical Engineering (Food

Technology) at Annamalai University. With five years of teaching experience, her research interests include food product development, functional foods, food packaging, and the sustainable conversion of food and agricultural by-products into value-added products. Her work emphasizes innovation, resource optimization, and environmentally responsible solutions for the food industry.

Sixth Author

Chandana C. is a Food Technologist from Kannur. She has completed her Bachelor of Technology (B.Tech.) degree from the Karpagam Academy of Higher Education and Master of Technology (M.Tech.) degree from the Jain (Deemed-to-be-University) in Food Technology. Her research focuses on sustainable food processing, food product development, value added food products and biodegradable solutions for food and food related applications.

Seventh Author

Mr. Lokanesh A. is a Subject Matter Expert (RCM) at GEBBS Healthcare Solutions with professional expertise in healthcare operations, process management, and quality-driven service delivery. He holds a Bachelor's degree in Chemical Engineering and possesses a strong interest in applying analytical and problem-solving approaches across interdisciplinary domains. His areas of interest include process optimization, operational excellence, and sustainable technology solutions. Through his professional contributions, he is committed to enhancing efficiency and innovation in both industrial and healthcare sectors.

CONTENTS

CHAPTER 1

INTRODUCTION TO WASTE VALORIZATION

1.1 Concept of Waste-to-Wealth

Waste-to-wealth is an innovative and sustainable approach that transforms waste materials into valuable products, energy sources, or industrial raw materials. Instead of treating waste as an unwanted byproduct, modern environmental engineering considers waste as a potential resource capable of generating economic, environmental, and social benefits.

Rapid industrialization, urbanization, and population growth have significantly increased the generation of agricultural, industrial, and municipal waste worldwide. Improper disposal of these wastes causes serious environmental issues such as land pollution, water contamination, greenhouse gas emissions, and depletion of natural resources. The waste-to-wealth concept aims to minimize these environmental impacts while simultaneously creating sustainable products and economic opportunities.

In recent years, agro-waste valorization has gained significant importance because agricultural residues contain valuable organic compounds such as cellulose, starch, lignin, pectin, proteins, and sugars that can be converted into bioplastics, biofuels, biomaterials, fertilizers, and decorative products.

Examples include:

- Production of PHA bioplastics from potato peel waste
- Development of PLA from rice processing wastewater

- Fabrication of vegan leather from agro-based polymers
- Preparation of decorative products from waste leaves and biomass fibers

The waste-to-wealth strategy supports sustainable industrial development and contributes to the reduction of environmental pollution through resource recovery and circular utilization of waste materials.

1.2 Circular Economy Principles

The circular economy is a sustainable economic model designed to reduce waste generation and maximize resource efficiency by maintaining materials within the production and consumption cycle for as long as possible. Unlike the traditional linear economy model based on "take–make–dispose," the circular economy focuses on reuse, recycling, recovery, remanufacturing, and regeneration.

The core principles of the circular economy include:

1. Reduce

Minimizing the consumption of raw materials, water, and energy during production processes.

2. Reuse

Reusing products and materials without significant modification to extend their service life.

3. Recycle

Processing waste materials into new products or raw materials for further industrial applications.

4. Recover

Recovering valuable compounds, nutrients, or energy from waste streams.

5. Regenerate

Restoring natural systems and encouraging the use of renewable and biodegradable materials.

In the context of agro-waste utilization, circular economy practices encourage:

- Conversion of food waste into biopolymers
- Use of biomass residues for sustainable products
- Composting and biodegradation of bioplastics
- Reduction of dependence on petroleum-based materials

The textile and packaging industries are increasingly adopting circular economy approaches to minimize environmental impacts and promote sustainable manufacturing systems.

1.3 Agro-Waste Generation and Environmental Issues

Agricultural activities generate enormous quantities of biomass waste every year. Agro-waste includes crop residues, fruit peels, leaves, husks, shells, stalks, and wastewater from food processing industries. Common examples include potato peel, rice husk, coconut leaves, sugarcane bagasse, banana fibers, and fruit processing residues.

Although agro-waste is biodegradable in nature, improper disposal can still create severe environmental problems.

Major Environmental Issues Associated with Agro-Waste

1. Open Burning

Farmers often burn agricultural residues to clear fields quickly, leading to:

- Air pollution
- Carbon dioxide emissions
- Release of particulate matter

- Respiratory health problems

2. Land Pollution

Accumulation of untreated biomass waste in landfills can:

- Produce foul odor
- Increase microbial contamination
- Occupy valuable land area

3. Water Pollution

Discharge of agro-industrial wastewater into water bodies can increase:

- Biological Oxygen Demand (BOD)
- Chemical Oxygen Demand (COD)
- Nutrient pollution

4. Greenhouse Gas Emissions

Decomposition of organic waste under anaerobic conditions releases methane, a potent greenhouse gas contributing to climate change.

Despite these challenges, agro-waste possesses significant potential for sustainable material production because of its rich organic composition. Proper utilization of agro-residues can reduce environmental burden while generating economic value.

1.4 Importance of Sustainable Materials

Sustainable materials are environmentally friendly materials produced from renewable, recyclable, or biodegradable resources with minimal ecological impact throughout their lifecycle.

The growing environmental concerns associated with synthetic plastics, fossil-fuel dependency, and industrial pollution have increased the demand for sustainable alternatives.

Characteristics of Sustainable Materials

- Renewable source origin
- Biodegradability
- Low toxicity
- Energy-efficient production
- Reduced carbon footprint
- Recyclability and compostability

Bioplastics such as PHA and PLA are emerging as promising sustainable materials due to their biodegradable nature and potential to replace conventional plastics in packaging, agriculture, textiles, and biomedical applications.

Similarly, bio-based vegan leather developed from agricultural waste offers a sustainable alternative to animal leather and petroleum-based synthetic leather.

Decorative products prepared from waste leaves and biomass fibers further demonstrate the potential of sustainable material engineering in reducing waste generation and promoting eco-friendly product development.

The adoption of sustainable materials supports:

- Environmental protection
- Resource conservation
- Circular economy implementation
- Sustainable industrial growth
- Green product innovation

1.5 Scope of Biomass Utilization

Biomass utilization refers to the conversion of organic biological materials into useful products, chemicals, fuels, and energy. Agricultural residues are considered one of the most abundant and renewable biomass resources available globally.

Biomass contains valuable components such as:

- Cellulose
- Hemicellulose
- Lignin
- Starch
- Sugars
- Natural fibers

These components can be processed into a wide variety of sustainable products.

Applications of Biomass Utilization

1. Bioplastic Production

Biomass-derived sugars and starches can be fermented to produce:

- Polyhydroxyalkanoates (PHA)
- Polylactic Acid (PLA)

2. Bio-Based Vegan Leather

Natural polymers extracted from agro-waste can be blended to produce biodegradable leather alternatives.

3. Bioenergy Generation

Biomass can be converted into:

- Biogas
- Bioethanol

- Biodiesel
- Biohydrogen

4. Decorative and Craft Products

Waste leaves, fibers, and agricultural residues can be transformed into:

- Flower vases
- Decorative panels
- Handicrafts
- Packaging materials

5. Soil and Agricultural Applications

Biomass residues can also be utilized for:

- Compost production
- Biofertilizers
- Soil conditioners

The future scope of biomass utilization is vast due to increasing interest in sustainable technologies, renewable resources, and eco-friendly industrial systems. Advancements in biotechnology, material science, and environmental engineering are expected to further enhance the commercial viability of biomass-based products.

The effective utilization of agro-waste not only reduces environmental pollution but also creates opportunities for innovation, employment generation, and sustainable economic development.

CHAPTER 2

AGRICULTURAL WASTE AS RENEWABLE RESOURCE

2.1 Introduction

Agricultural activities generate enormous quantities of biomass residues every year across the world. These wastes are produced during cultivation, harvesting, food processing, transportation, and storage operations. Traditionally, agricultural waste materials were considered low-value residues and were often disposed of through open burning, dumping, or landfilling, leading to severe environmental pollution and greenhouse gas emissions.

In recent years, agricultural waste has gained significant attention as a renewable resource due to its potential for producing sustainable materials, biofuels, bioplastics, bio-composites, and other value-added products. Agro-waste contains valuable organic compounds such as cellulose, starch, lignin, proteins, fibers, and sugars that can be utilized for industrial applications.

The concept of converting agricultural waste into useful products supports:

- Waste valorization
- Circular economy principles
- Sustainable industrial development
- Renewable resource utilization

This chapter discusses the types of agro-waste, characteristics of potato peel waste, rice water and food-processing waste, coconut leaves and biomass residues, and the industrial potential of agricultural waste resources.

2.2 Types of Agro-Waste

Agricultural waste refers to residues generated from farming activities, food processing industries, and agro-based operations. These wastes can be classified based on their origin and composition.

2.2.1 Field Residues

Field residues are generated during harvesting and agricultural operations.

Examples

- Rice straw
- Wheat straw
- Corn stalks
- Sugarcane leaves
- Coconut leaves

Characteristics

- High lignocellulosic content
- Abundant availability
- Seasonal generation

Applications

- Biofuel production
- Fiber extraction
- Biocomposites
- Decorative products

2.2.2 Processing Residues

Processing residues are generated during food and agricultural product processing.

Examples

- Potato peel
- Fruit peels
- Rice washing water
- Sugarcane bagasse
- Coffee husk

Characteristics

- Rich in starch and sugars
- High organic content
- Biodegradable nature

Applications

- Bioplastic production
- Fermentation substrates
- Animal feed
- Composting

2.2.3 Animal-Based Agricultural Waste

Generated from livestock and poultry industries.

Examples

- Manure
- Poultry litter
- Animal residues

Applications

- Biogas production
- Organic fertilizer
- Soil conditioning

2.2.4 Forestry and Plantation Residues

Generated from plantation and forestry activities.

Examples

- Coconut shells
- Palm leaves
- Wood chips
- Bark residues

Applications

- Activated carbon
- Decorative products
- Biomass energy

2.3 Potato Peel Waste Characteristics

2.3.1 Introduction

Potato is one of the most widely consumed food crops worldwide. Food processing industries involved in chip manufacturing, frozen foods, and starch production generate large quantities of potato peel waste.

Potato peel accounts for approximately 15–40% of total potato processing waste depending on peeling methods and industrial operations.

Improper disposal of potato peel waste can cause:

- Odor generation

- Microbial growth
- Methane emission
- Environmental pollution

However, potato peel contains valuable biochemical compounds suitable for sustainable material production.

2.3.2 Composition of Potato Peel

Potato peel contains:

- Starch
- Cellulose
- Hemicellulose
- Proteins
- Fibers
- Minerals

Typical Composition

Component	Approximate Percentage
Moisture	70–80%
Starch	20–30%
Fiber	10–15%
Protein	5–10%
Ash	3–5%

2.3.3 Important Characteristics

1. High Starch Content

Potato peel is rich in carbohydrates, making it suitable for:

- Fermentation
- Bioplastic production
- Bioethanol generation

2. Biodegradable Nature

Being organic biomass, potato peel decomposes naturally under environmental conditions.

3. Renewable Resource

Continuous agricultural production ensures abundant availability.

4. Fermentable Sugar Source

Hydrolysis of starch produces sugars useful for:

- PHA production
- Lactic acid fermentation

2.3.4 Industrial Potential

Potato peel waste can be utilized for:

- PHA bioplastic production
- Animal feed
- Biofuel production
- Compost generation
- Starch extraction

2.4 Rice Water and Food Processing Waste

2.4.1 Introduction

Rice washing and cooking processes generate large quantities of rice water rich in starch and dissolved nutrients. Food processing industries also generate wastewater and organic residues that possess significant industrial potential.

Traditionally, rice water is discarded directly into drainage systems, contributing to:

- Water pollution
- Increased Biological Oxygen Demand (BOD)
- Nutrient accumulation

However, rice water contains carbohydrates and nutrients suitable for biotechnological applications

2.4.2 Characteristics of Rice Water

Rice water contains:

- Starch
- Sugars
- Proteins
- Vitamins
- Minerals

Important Properties

- High carbohydrate concentration
- Biodegradable nature
- Fermentation suitability

2.4.3 Applications of Rice Water

- Lactic acid production
- PLA bioplastic production
- Biofertilizer development
- Fermentation media

2.4.4 Food Processing Waste

Food industries generate various waste materials such as:

- Fruit peels
- Vegetable residues
- Wastewater
- Starch-rich residues

These wastes can be converted into:

- Bioplastics
- Bioenergy
- Animal feed
- Organic fertilizers

2.4.5 Environmental Concerns

Improper disposal causes:

- Water contamination
- Odor problems
- Greenhouse gas emissions

Sustainable utilization helps minimize environmental impact.

2.5 Coconut Leaves and Biomass Residues

2.5.1 Introduction

Coconut cultivation generates large quantities of leaf and shell waste. Coconut leaves are often discarded or burned after harvesting, resulting in biomass waste accumulation.

- However, coconut leaves possess strong fibrous structures suitable for various sustainable applications.

2.5.2 Characteristics of Coconut Leaves

Physical Characteristics

- Fibrous texture
- Lightweight structure
- Flexibility
- Natural durability

Chemical Composition

- Cellulose
- Hemicellulose
- Lignin

2.5.3 Applications of Coconut Leaves

- Decorative products
- Fiber extraction
- Handmade crafts
- Biocomposites
- Eco-friendly packaging

2.5.4 Other Biomass Residues

Examples

- Banana fibers
- Pineapple leaves
- Sugarcane bagasse
- Corn husk
- Palm leaves

2.5.5 Industrial Importance

Biomass residues can be used for:

- Natural fiber composites
- Biofuel production
- Paper manufacturing
- Biodegradable materials

2.6 Availability and Industrial Potential of Agro-Waste

2.6.1 Availability of Agro-Waste

Agricultural waste is abundantly available due to:

- Large-scale farming activities
- Food processing industries
- Plantation agriculture
- Seasonal crop production

Developing countries generate enormous quantities of agro-waste annually.

2.6.2 Industrial Potential

Agro-waste possesses immense industrial potential due to its renewable and biodegradable nature.

2.6.3 Applications in Bioplastic Industry

Agro-waste can serve as:

- Fermentation substrate
- Starch source
- Polymer precursor

for:

- PHA production
- PLA synthesis
- Starch-based plastics

2.6.4 Applications in Textile Industry

Natural fibers extracted from agro-waste are used for:

- Vegan leather
- Sustainable textiles
- Composite materials
- Eco-fashion products

2.6.5 Applications in Packaging Industry

Biomass-derived materials are used for:

- Biodegradable films
- Eco-friendly containers
- Cushion packaging
- Compostable trays

2.6.6 Applications in Decorative Product Engineering

Agricultural residues are utilized for:

- Flower vases
- Wall décor
- Handicrafts
- Sustainable home décor products

2.6.7 Economic Importance

Agro-waste utilization supports:

- Rural employment
- Waste management industries
- Sustainable entrepreneurship
- Value-added product development

2.7 Environmental Benefits of Agro-Waste Utilization

Utilization of agricultural waste provides several environmental advantages.

Major Benefits

- Reduction in landfill waste
- Lower greenhouse gas emissions
- Prevention of open burning
- Reduced environmental pollution
- Sustainable resource utilization

Agro-waste valorization also supports circular economy systems by converting waste into valuable industrial products.

2.8 Challenges in Agro-Waste Utilization

Despite its advantages, agro-waste utilization faces several challenges.

Major Challenges

- Seasonal availability
- Moisture content variation
- Storage difficulties

- Transportation cost
- Lack of industrial infrastructure

Research and technological advancements are required to improve processing efficiency and commercialization.

2.9 Future Scope

The future of agro-waste utilization is highly promising due to:

- Increasing environmental awareness
- Demand for sustainable materials
- Circular economy implementation
- Advancement in green technologies

Future applications may include:

- Smart biomaterials
- Nano-biocomposites
- Sustainable textiles
- Green packaging systems

2.10 Conclusion

Agricultural waste represents a valuable renewable resource with immense potential for sustainable industrial applications. Waste materials such as potato peel, rice water, coconut leaves, and other biomass residues contain useful organic compounds suitable for bioplastic production, vegan leather fabrication, decorative product engineering, and bio-based industrial materials.

Proper utilization of agro-waste can significantly reduce environmental pollution, improve resource efficiency, and support circular economy systems. The conversion of agricultural residues into value-added products offers promising opportunities for sustainable development, green entrepreneurship, and future environmentally responsible industries.

CHAPTER 3

FUNDAMENTALS OF BIOPLASTICS

3.1 Introduction

Plastics have become an essential part of modern society because of their versatility, durability, lightweight nature, and low production cost. They are extensively used in packaging, construction, automotive, textile, healthcare, agriculture, and consumer products. However, the excessive use of conventional petroleum-based plastics has created severe environmental concerns due to their non-biodegradable nature and accumulation in ecosystems.

The increasing awareness of plastic pollution, climate change, and fossil fuel depletion has accelerated research toward sustainable alternatives known as bioplastics. Bioplastics are environmentally friendly materials derived from renewable biological resources and are often biodegradable or compostable under suitable conditions.

Among various bioplastics, Polyhydroxyalkanoates (PHA) and Polylactic Acid (PLA) are considered promising materials because of their biodegradability, renewable origin, and broad industrial applications.

This chapter discusses the fundamentals of bioplastics, differences between conventional plastics and bioplastics, major types of bioplastics, properties of PHA and PLA, and the concepts of biodegradability and compostability.

3.2 Conventional Plastics vs Bioplastics

3.2.1 Conventional Plastics

Conventional plastics are synthetic polymers produced mainly from petroleum and natural gas. Common conventional plastics include:

- Polyethylene (PE)
- Polypropylene (PP)
- Polyvinyl Chloride (PVC)
- Polystyrene (PS)
- Polyethylene Terephthalate (PET)

These plastics possess excellent durability, chemical resistance, and mechanical strength, making them suitable for long-term applications. However, their resistance to degradation causes long-term environmental pollution.

Advantages of Conventional Plastics

- High durability
- Low production cost
- Good mechanical properties
- Water resistance
- Easy processability

Disadvantages of Conventional Plastics

- Non-biodegradable
- Derived from fossil fuels
- Microplastic generation
- Greenhouse gas emissions
- Land and marine pollution

Improper disposal of conventional plastics results in accumulation in landfills and aquatic ecosystems, affecting wildlife and human health.

3.2.2 Bioplastics

Bioplastics are polymers produced from renewable biological resources such as starch, cellulose, sugars, vegetable oils, and agricultural waste. Some bioplastics are biodegradable, while others are bio-based but non-biodegradable.

Raw Materials Used for Bioplastic Production

- Potato peel
- Corn starch
- Sugarcane bagasse
- Rice wastewater
- Fruit peels
- Vegetable oils
- Cellulosic biomass

Advantages of Bioplastics

- Renewable resource origin
- Reduced carbon footprint
- Biodegradability
- Compostability
- Lower environmental impact

Limitations of Bioplastics

- Higher production cost
- Lower thermal stability in some cases
- Limited large-scale industrial infrastructure
- Specific composting conditions required

Bioplastics are increasingly used in packaging, biomedical devices, textiles, agriculture, disposable products, and automotive components.

3.3 Types of Bioplastics

Bioplastics are classified based on their source materials and biodegradability characteristics.

3.3.1 Starch-Based Bioplastics

These bioplastics are produced using starch extracted from:

- Corn
- Potato
- Tapioca
- Wheat

Starch-based plastics are biodegradable and commonly used in packaging applications.

Advantages

- Low cost
- Easily biodegradable
- Renewable source

Limitations

- Poor moisture resistance
- Lower mechanical strength

3.3.2 Cellulose-Based Bioplastics

Cellulose is the most abundant natural polymer found in plants. Cellulose derivatives are used to produce biodegradable films and coatings.

Applications

- Packaging films
- Textile coatings
- Biomedical materials

3.3.3 Protein-Based Bioplastics

Proteins such as soy protein, gelatin, and casein are used for producing biodegradable materials.

Characteristics

- Good film-forming ability
- Biocompatibility
- Moderate mechanical strength

3.3.4 Polylactic Acid (PLA)

PLA is a thermoplastic polyester produced from lactic acid obtained through fermentation of sugars and starch-rich biomass.

Major Features

- Transparent appearance
- Good mechanical properties
- Biodegradable under industrial composting conditions

Applications

- Food packaging
- Disposable cups
- 3D printing
- Biomedical implants
- Textile fibers

3.3.5 Polyhydroxyalkanoates (PHA)

PHA is a family of biodegradable polyesters synthesized naturally by microorganisms as intracellular energy storage compounds.

Key Features

- Fully biodegradable
- Biocompatible
- Thermoplastic behavior

Applications

- Packaging films
- Medical devices
- Agricultural mulch films
- Biodegradable containers

3.4 Polyhydroxyalkanoates (PHA)

3.4.1 Introduction to PHA

Polyhydroxyalkanoates (PHA) are biodegradable polymers synthesized by bacteria under nutrient-limited conditions with excess carbon sources. PHA acts as an intracellular energy reserve for microorganisms.

Common bacterial strains used for PHA production include:

- Cupriavidus necator
- Bacillus species
- Pseudomonas species

Agro-waste materials such as potato peel, fruit waste, and food-processing wastewater can serve as economical substrates for microbial fermentation.

3.4.2 Production of PHA from Potato Peel

Potato peel contains high starch content, making it a suitable substrate for microbial fermentation.

Production Steps

1. Collection and cleaning of potato peel
2. Drying and grinding
3. Hydrolysis to release fermentable sugars
4. Fermentation using bacterial cultures
5. PHA accumulation inside cells
6. Extraction and purification

3.4.3 Properties of PHA

- Biodegradable
- Biocompatible
- Water-insoluble
- Thermoplastic nature
- Good oxygen barrier properties

3.4.4 Applications of PHA

- Food packaging
- Biomedical implants
- Drug delivery systems
- Disposable products
- Agricultural films

PHA is considered one of the most environmentally sustainable alternatives to petroleum-based plastics.

3.5 Polylactic Acid (PLA)

3.5.1 Introduction to PLA

Polylactic Acid (PLA) is one of the most commercially successful bioplastics. It is produced through fermentation of renewable carbohydrate sources into lactic acid, followed by polymerization.

Sources for PLA Production

- Corn starch
- Sugarcane
- Rice processing wastewater
- Agricultural biomass

3.5.2 PLA Production Process

Step 1: Fermentation

Sugars are fermented using microorganisms to produce lactic acid.

Step 2: Purification

The produced lactic acid is purified.

Step 3: Polymerization

Lactic acid molecules undergo polymerization to form PLA.

3.5.3 Properties of PLA

- Transparent appearance
- High tensile strength
- Good processability
- Compostable under industrial conditions
- Low toxicity

3.5.4 Applications of PLA

- Packaging materials
- Disposable tableware
- Textile fibers
- Biomedical scaffolds
- 3D printing filaments

PLA is increasingly used as a sustainable alternative in the packaging and textile industries.

3.6 Biodegradability and Compostability

3.6.1 Biodegradability

Biodegradability refers to the ability of a material to decompose naturally through the action of microorganisms such as bacteria, fungi, and algae.

During biodegradation:

- Complex polymers break into smaller molecules
- Final products include carbon dioxide, water, methane, and biomass

Biodegradable plastics help reduce long-term environmental pollution compared to conventional plastics.

Factors Affecting Biodegradation

- Temperature
- Moisture content
- Microbial activity
- Oxygen availability
- Material composition

3.6.2 Compostability

Compostability is the ability of a material to biodegrade under controlled composting conditions into non-toxic organic matter.

Industrial composting typically requires:

- Elevated temperature
- Controlled humidity
- Aerobic conditions

Characteristics of Compostable Plastic

- Rapid degradation
- Non-toxic residue formation
- No harmful microplastics

3.6.3 Importance of Biodegradable and Compostable Materials

- Reduction in plastic waste accumulation
- Lower environmental pollution
- Support for circular economy
- Sustainable waste management
- Reduced fossil fuel dependency

Bioplastics such as PHA and PLA contribute significantly toward sustainable development goals by promoting renewable resource utilization and environmentally responsible material production.

3.7 Conclusion

Bioplastics represent an important advancement toward sustainable material development and environmental protection. Unlike conventional petroleum-based plastics, bioplastics are derived from renewable biomass resources and offer biodegradability and compostability advantages.

Among the different bioplastics, PHA and PLA have gained significant attention because of their excellent properties and wide industrial applications. Agro-waste materials such as potato peel and rice wastewater provide sustainable feedstocks for bioplastic production while simultaneously addressing waste management challenges.

The development and commercialization of biodegradable bioplastics can play a major role in reducing plastic pollution, supporting circular economy practices, and promoting sustainable industrial growth in the future.

CHAPTER 4

PRODUCTION OF PHA BIOPLASTIC FROM POTATO PEEL

4.1 Introduction

The growing environmental concerns associated with conventional petroleum-based plastics have accelerated the development of biodegradable and sustainable alternatives. Among various bioplastics, Polyhydroxyalkanoates (PHA) have emerged as promising eco-friendly polymers due to their biodegradability, biocompatibility, and renewable origin.

PHA is a microbial polyester synthesized naturally by various bacteria under nutrient-limited conditions with excess carbon sources. Agricultural and food-processing wastes are increasingly being explored as economical substrates for PHA production to reduce production costs and support waste valorization strategies.

Potato peel waste is one of the most suitable agro-industrial residues for PHA production because of its high starch and carbohydrate content. Large quantities of potato peel are generated from food processing industries, restaurants, and households, and improper disposal can create environmental problems. Converting potato peel into PHA bioplastics offers a sustainable waste-to-wealth approach while simultaneously reducing organic waste accumulation.

This chapter discusses the composition of potato peel, pretreatment methods, fermentation techniques, microorganisms used for PHA synthesis, extraction and purification methods, factors affecting PHA yield, and various applications of PHA bioplastics.

4.2 Composition of Potato Peel

Potato peel is a nutrient-rich agricultural waste generated during potato processing operations such as peeling, slicing, and chip manufacturing. It contains significant amounts of carbohydrates and organic compounds suitable for microbial fermentation.

4.2.1 Chemical Composition of Potato Peel

Component	Approximate Composition (%)
Starch	40–60
Cellulose	10–20
Hemicellulose	8–15
Lignin	2–5
Proteins	5–10
Moisture	70–80
Minerals and Ash	3–5

The high starch content makes potato peel an excellent carbon source for microbial growth and PHA accumulation.

4.2.2 Advantages of Potato Peel as Feedstock

- Easily available agro-waste
- Low-cost substrate
- Rich in fermentable carbohydrates
- Renewable and biodegradable
- Reduces food processing waste disposal problems

Potato peel waste can therefore serve as a sustainable raw material for bioplastic production.

4.3 Pretreatment Methods

Pretreatment is an essential step to convert complex carbohydrates present in potato peel into simple fermentable sugars that microorganisms can utilize during fermentation.

The efficiency of pretreatment significantly influences microbial growth and PHA production.

4.3.1 Objectives of Pretreatment

- Removal of impurities
- Reduction of microbial contamination
- Breakdown of starch and lignocellulosic structure
- Release of fermentable sugars
- Improvement of substrate accessibility

4.3.2 Physical Pretreatment

a) Washing

Potato peels are washed thoroughly to remove:

- Dirt
- Soil particles
- Oil residues
- Other contaminants

b) Drying

The washed peels are dried using:

- Sun drying

- Hot air oven drying

Drying reduces moisture content and prevents microbial spoilage.

c) Grinding

Dried potato peels are ground into powder form to increase surface area and improve hydrolysis efficiency.

4.3.3 Chemical Pretreatment

Chemical pretreatment involves the use of acids or alkalis to hydrolyze starch and lignocellulosic components.

Acid Hydrolysis

Common acids used:

- Hydrochloric acid (HCl)
- Sulfuric acid (H_2SO_4)

Acid treatment converts starch into glucose and other fermentable sugars.

Advantages

- Rapid hydrolysis
- High sugar release

Limitations

- Corrosive nature
- Formation of inhibitory compounds

4.3.4 Enzymatic Pretreatment

Enzymes such as:

- Amylase
- Cellulase

- Glucoamylase

are used to hydrolyze starch and cellulose under mild conditions.

Advantages

- Environmentally friendly
- High specificity
- Reduced inhibitor formation

Limitations

- Higher enzyme cost
- Longer processing time

4.4 Fermentation Techniques

Fermentation is the biological conversion of organic substrates into valuable products using microorganisms. In PHA production, fermentation enables bacteria to accumulate intracellular polymer granules

4.4.1 Principle of PHA Fermentation

PHA synthesis occurs when microorganisms experience:

- Excess carbon supply
- Limitation of essential nutrients such as nitrogen or phosphorus

Under these stress conditions, microorganisms store excess carbon as PHA.

4.4.2 Batch Fermentation

In batch fermentation:

- All nutrients are added initially
- Fermentation proceeds in a closed system

Advantages

- Simple operation
- Low contamination risk

Limitations

- Lower productivity
- Nutrient depletion over time

4.4.3 Fed-Batch Fermentation

Substrate is added continuously or intermittently during fermentation.

Advantages

- Better substrate control
- Higher cell density
- Improved PHA yield

Fed-batch fermentation is commonly used in industrial PHA production.

4.4.4 Continuous Fermentation

Fresh medium is continuously added while culture broth is removed simultaneously.

Advantages

- High productivity
- Continuous operation

Limitations

- Higher contamination risk
- Complex operation

4.5 Microorganisms Used for PHA Production

Several microorganisms can synthesize PHA intracellularly under suitable environmental conditions.

4.5.1 Bacterial Species Used

1. Cupriavidus necator

One of the most widely studied PHA-producing bacteria.

Features:

- High PHA accumulation capacity
- Efficient utilization of sugars

2. Bacillus species

Gram-positive bacteria capable of producing PHA from agricultural waste substrates.

Advantages:

- Easy cultivation
- Fast growth rat

3. Pseudomonas species

Used mainly for medium-chain-length PHA production.

Applications:

- Flexible biodegradable plastics
- Biomedical materials

4. Alcaligenes species

Efficient PHA producers under nutrient stress conditions.

4.5.2 Factors Influencing Microbial PHA Production

- Carbon source concentration
- Nitrogen limitation

- Temperature
- pH
- Aeration
- Fermentation time

4.6 Extraction and Purification of PHA

After fermentation, PHA must be extracted from microbial cells and purified for further applications.

4.6.1 Cell Harvesting

Microbial cells are separated from fermentation broth using:

- Centrifugation
- Filtration

4.6.2 Cell Disruption Methods

Physical Methods

- Ultrasonication
- High-pressure homogenization
- Mechanical grinding

Chemical Methods

- Sodium hypochlorite digestion
- Solvent treatment

4.6.3 Solvent Extraction

Common solvents used:

- Chloroform
- Ethyl acetate

- Dichloromethane

PHA dissolves in solvents and is later precipitated using alcohol.

4.6.4 Purification

Purification removes:

- Cell debris
- Proteins
- Lipids
- Residual impurities

Purified PHA is then dried and converted into pellets or films.

4.7 Factors Affecting Yield

The yield of PHA production depends on several biological and process parameters.

4.7.1 Carbon Source

High carbohydrate concentration improves microbial growth and PHA accumulation.

Potato peel provides:

- Glucose
- Starch
- Organic nutrients

4.7.2 Nutrient Limitation

Nitrogen or phosphorus limitation promotes intracellular PHA storage.

4.7.3 pH

Optimal pH range:

- 6.5–7.5 for most bacteria

Improper pH reduces microbial activity.

4.7.4 Temperature

Most PHA-producing bacteria grow efficiently between:

- 30–37°C

4.7.5 Aeration and Agitation

Adequate oxygen supply improves:

- Cell growth
- Metabolic activity
- PHA accumulation

4.7.6 Fermentation Time

Longer fermentation increases PHA accumulation up to an optimum level.

Excessive fermentation may lead to:

- Nutrient depletion
- Polymer degradation

4.8 Applications of PHA Bioplastics

PHA possesses excellent biodegradability and biocompatibility, making it suitable for numerous industrial applications.

4.8.1 Packaging Industry

PHA films and containers are used as biodegradable alternatives to petroleum plastics.

Applications:

- Food packaging
- Disposable cups
- Carry bags

4.8.2 Biomedical Applications

PHA is biocompatible and suitable for medical applications.

Uses:

- Drug delivery systems
- Tissue engineering
- Surgical sutures
- Orthopedic implants

4.8.3 Agricultural Applications

PHA-based biodegradable films are used in:

- Mulching films
- Seed coatings
- Controlled fertilizer release systems

4.8.4 Textile and Fashion Industry

PHA can be incorporated into:

- Biodegradable fibers
- Eco-friendly leather alternatives
- Sustainable packaging material

4.8.5 Consumer Products

PHA is used in:

- Disposable cutlery

- Cosmetic packaging
- Household items

4.9 Environmental Benefits of PHA

PHA contributes significantly toward environmental sustainability because:

- It is biodegradable
- Derived from renewable resources
- Reduces plastic pollution
- Lowers greenhouse gas emissions
- Supports circular economy practices

Utilizing potato peel waste for PHA production also reduces food-processing waste disposal problems and promotes agro-waste valorization.

4.10 Conclusion

PHA bioplastics produced from potato peel waste represent a sustainable and eco-friendly alternative to conventional plastics. Potato peel contains abundant starch and fermentable carbohydrates suitable for microbial fermentation and polymer synthesis.

The production process involves pretreatment, fermentation, microbial accumulation, extraction, and purification steps. Various microorganisms such as *Cupriavidus necator* and *Bacillus* species can efficiently synthesize PHA under nutrient-limited conditions.

PHA offers excellent biodegradability, biocompatibility, and industrial applicability in packaging, biomedical, agricultural, and textile sectors. The conversion of potato peel waste into biodegradable plastics demonstrates an effective waste-to-wealth strategy supporting circular economy principles and sustainable industrial development.

CHAPTER 5

PLA AND BIO-BASED POLYMER DEVELOPMENT

5.1 Introduction

The increasing environmental concerns associated with conventional petroleum-based plastics have accelerated the development of sustainable and biodegradable polymeric materials. Among various bio-based polymers, Polylactic Acid (PLA) has emerged as one of the most promising biodegradable thermoplastics due to its renewable origin, good mechanical properties, and industrial applicability.

PLA is produced from lactic acid obtained through the fermentation of renewable biomass resources such as starch, sugars, and agro-industrial waste materials. The use of agro-waste for PLA production not only reduces production costs but also promotes sustainable waste management and circular economy practices.

Agricultural and food-processing wastes such as rice wastewater, potato waste, sugarcane bagasse, corn starch residues, and fruit peels are increasingly explored as economical substrates for lactic acid production. The development of PLA from renewable resources offers a sustainable alternative to conventional plastics in packaging, textiles, biomedical devices, and consumer products.

This chapter discusses PLA production from agro-waste, lactic acid fermentation, polymerization techniques, properties of PLA, and its industrial applications.

5.2 PLA Production from Agro-Waste

5.2.1 Agro-Waste as Feedstock for PLA Production

Agro-waste contains abundant carbohydrates, sugars, starch, cellulose, and organic compounds that can be converted into lactic acid through microbial fermentation.

Common agro-waste materials used for PLA production include:

- Rice processing wastewater
- Potato peel waste
- Corn starch residues
- Sugarcane bagasse
- Fruit peels
- Cassava waste
- Food-processing wastewater

These wastes provide low-cost renewable substrates for bio-based polymer production.

5.2.2 Advantages of Agro-Waste Utilization

- Reduction of agricultural waste disposal problems
- Lower production cost
- Renewable and sustainable feedstock
- Reduction of fossil fuel dependency
- Support for circular economy principles
- Reduction in greenhouse gas emissions

The conversion of agro-waste into PLA demonstrates an effective waste-to-wealth strategy.

5.2.3 Steps in PLA Production from Agro-Waste

1. Collection of Agro-Waste

Agricultural residues and food-processing wastes are collected from industries and agricultural fields.

2. Pretreatment

Pretreatment converts complex carbohydrates into fermentable sugars through:

- Physical treatment
- Chemical hydrolysis
- Enzymatic hydrolysis

3. Fermentation

Microorganisms ferment sugars to produce lactic acid.

4. Purification

Lactic acid is purified to remove impurities and by-products.

5. Polymerization

Purified lactic acid undergoes polymerization to form PLA.

6. Product Processing

PLA is converted into:

- Films
- Fibers
- Sheets
- Pellets
- Molded products

5.3 Lactic Acid Fermentation

5.3.1 Introduction to Lactic Acid Fermentation

Lactic acid fermentation is a biological process in which microorganisms convert carbohydrates into lactic acid under controlled environmental conditions.

Lactic acid is the primary monomer required for PLA synthesis.

5.3.2 Microorganisms Used for Lactic Acid Production

Several microorganisms are capable of producing lactic acid efficiently.

Common Microorganisms

- *Lactobacillus species*
- *Bacillus coagulans*
- *Rhizopus oryzae*
- *Streptococcus thermophilus*

Among these, *Lactobacillus* species are most commonly used due to their high lactic acid productivity.

5.3.3 Fermentation Process

Step 1: Preparation of Fermentation Medium

Hydrolyzed agro-waste containing fermentable sugars is prepared as substrate.

Step 2: Sterilization

The medium is sterilized to prevent contamination.

Step 3: Inoculation

Selected microbial culture is introduced into the fermentation medium.

Step 4: Fermentation

Microorganisms convert sugars into lactic acid under controlled conditions.

Step 5: Recovery

Lactic acid is separated and purified.

5.3.4 Factors Affecting Lactic Acid Fermentation

1. pH

Optimal pH range:

- 5.5–6.5

Improper pH reduces microbial growth.

2. Temperature

Optimal temperature:

- 30–45°C depending on microorganism

3. Sugar Concentration

Adequate carbohydrate concentration enhances lactic acid yield.

4. Oxygen Availability

Most lactic acid bacteria prefer anaerobic or microaerophilic conditions.

5. Fermentation Time

Longer fermentation increases lactic acid production up to an optimum level.

5.4 Polymerization Techniques

Polymerization converts lactic acid monomers into long-chain PLA polymers.

Several polymerization methods are used for PLA synthesis.

5.4.1 Direct Polycondensation

In direct polycondensation:

- Lactic acid molecules combine directly
- Water is removed during the reaction

Advantages

- Simple process
- Low equipment requirement

Limitations

- Lower molecular weight PLA
- Longer reaction time

5.4.2 Ring-Opening Polymerization (ROP)

Ring-opening polymerization is the most widely used industrial technique for PLA production.

Process

- Lactic acid forms cyclic lactide intermediates
- Lactide undergoes polymerization using catalysts

Catalysts Used

- Tin octoate
- Zinc compounds
- Aluminum catalysts

Advantages

- High molecular weight PLA
- Better mechanical properties
- Industrial scalability

5.4.3 Azeotropic Dehydration Polymerization

This method removes water continuously using azeotropic solvents during polymerization.

Advantages

- Improved polymer quality
- Higher molecular weight

Limitations

- Complex process
- Higher cost

5.5 Properties of PLA

PLA possesses several desirable physical, chemical, and mechanical properties suitable for industrial applications.

5.5.1 Physical Properties

- Transparent appearance
- Glossy surface
- Lightweight material
- Thermoplastic nature

5.5.2 Mechanical Properties

- High tensile strength
- Good stiffness
- Moderate flexibility
- Good dimensional stability

5.5.3 Thermal Properties

- Melting temperature: 150–180°C
- Glass transition temperature: 55–65°C

PLA exhibits good thermal processability but limited heat resistance compared to some conventional plastics.

5.5.4 Biodegradability

PLA can degrade into:

- Carbon dioxide
- Water
- Biomass

under industrial composting conditions.

5.5.5 Biocompatibility

PLA is non-toxic and biocompatible, making it suitable for biomedical applications.

5.5.6 Barrier Properties

PLA provides moderate barrier properties against:

- Oxygen
- Moisture
- Aroma compounds

5.6 Industrial Applications of PLA

Due to its biodegradability and renewable origin, PLA is widely used in several industries.

5.6.1 Packaging Industry

PLA is extensively used for:

- Food packaging films
- Disposable cups
- Bottles
- Containers
- Carry bags

Advantages

- Compostable packaging
- Reduced plastic pollution

5.6.2 Textile Industry

PLA fibers are used in:

- Sustainable fabrics
- Nonwoven textiles
- Upholstery materials
- Fashion accessories

PLA-based textiles offer eco-friendly alternatives to synthetic fibers.

5.6.3 Biomedical Applications

Because of its biocompatibility, PLA is used in:

- Surgical sutures
- Drug delivery systems
- Tissue engineering scaffolds
- Orthopedic implants

PLA degrades naturally inside the body over time.

5.6.4 3D Printing Industry

PLA is one of the most commonly used materials in additive manufacturing.

Advantages in 3D Printing

- Easy processability
- Low warping
- Low toxicity

Applications include:

- Prototypes
- Educational models
- Consumer products

5.6.5 Agricultural Applications

PLA is used in:

- Biodegradable mulch films
- Plant pots
- Controlled-release systems

5.6.6 Automotive Industry

PLA composites are utilized in:

- Interior panels
- Decorative components
- Lightweight automotive materials

5.7 Environmental Benefits of PLA

PLA contributes significantly toward sustainable development.

Major Environmental Benefits

- Renewable raw material source
- Reduction in fossil fuel usage
- Lower greenhouse gas emissions
- Biodegradability
- Reduced landfill burden
- Support for circular economy systems

Utilization of agro-waste for PLA production further enhances environmental sustainability by converting waste into valuable bio-based products.

5.8 Challenges in PLA Development

Despite its advantages, PLA production faces several challenges.

Major Challenges

- High production cost
- Limited thermal resistance
- Industrial composting requirement
- Brittleness in some applications
- Need for process optimization

Research is ongoing to improve PLA properties using:

- Plasticizers
- Natural fibers
- Nanocomposites
- Polymer blending

5.9 Future Scope of PLA Development

The future of PLA development is highly promising due to increasing environmental regulations and demand for sustainable materials.

Emerging areas include:

- Smart biodegradable materials
- Nano-enhanced PLA composites
- Sustainable textile applications
- Bio-based packaging systems
- Medical biomaterials
- Circular economy integration

Advancements in biotechnology and fermentation engineering are expected to reduce production costs and improve commercial viability.

5.10 Conclusion

Polylactic Acid (PLA) is one of the most important biodegradable biopolymers developed from renewable biomass resources. Agro-waste materials such as rice wastewater, potato residues, and food-processing wastes provide sustainable substrates for lactic acid fermentation and PLA synthesis.

The production of PLA involves pretreatment, fermentation, purification, and polymerization processes. Among the polymerization methods, ring-opening polymerization is widely used for industrial-scale production due to its ability to produce high-quality polymers.

PLA possesses excellent biodegradability, biocompatibility, and mechanical properties suitable for packaging, textiles, biomedical devices, and 3D printing applications. The development of PLA from agro-waste represents an effective strategy for sustainable material production, environmental protection, and circular economy implementation.

CHAPTER 6

SUSTAINABLE VEGAN LEATHER FABRICATION

6.1 Introduction

The increasing environmental concerns associated with conventional leather production and synthetic petroleum-based leather materials have accelerated the development of sustainable and eco-friendly alternatives known as vegan leather. Vegan leather refers to leather-like materials produced without the use of animal hides and with minimal environmental impact.

Conventional leather manufacturing involves extensive consumption of water, chemicals, energy, and animal resources. The tanning process also generates hazardous wastes and toxic effluents that contribute to environmental pollution. Similarly, synthetic leather made from polyvinyl chloride (PVC) and polyurethane (PU) relies heavily on fossil fuels and creates non-biodegradable waste.

Sustainable vegan leather fabricated from agricultural waste and bio-based polymers offers an environmentally responsible alternative to traditional leather materials. Agro-waste materials such as fruit peels, rice water, potato peel, pineapple leaves, banana fibers, coconut residues, and pectin-rich biomass can be converted into biodegradable leather-like sheets suitable for textile and fashion applications.

This chapter discusses the need for vegan leather, environmental impacts of conventional leather production, agro-waste-based vegan leather development, bio-pectin extraction, composite sheet preparation, material properties, and textile applications.

6.2 Need for Vegan Leather

The global demand for sustainable materials has increased significantly due to environmental awareness, ethical concerns, and stricter environmental regulations. The fashion and textile industries are among the largest contributors to pollution, resource depletion, and waste generation.

Vegan leather has emerged as a sustainable alternative because it eliminates dependence on animal hides and reduces environmental burden.

6.2.1 Ethical Concerns

Traditional leather production involves the use of animal skin obtained from:

- Cattle
- Goats
- Sheep
- Buffalo

Growing awareness regarding animal welfare has increased consumer demand for cruelty-free materials.

6.2.2 Environmental Sustainability

Vegan leather supports:

- Reduced greenhouse gas emissions
- Lower water consumption
- Reduced chemical pollution
- Sustainable resource utilization

6.2.3 Demand in Fashion Industry

Modern consumers increasingly prefer:

- Eco-friendly products
- Biodegradable materials
- Sustainable fashion accessories

Leading fashion industries are therefore exploring plant-based leather alternatives.

6.2.4 Circular Economy Integration

Agro-waste-based vegan leather promotes:

- Waste valorization
- Resource recovery
- Circular material flow
- Sustainable production systems

6.3 Conventional Leather Environmental Impact

Conventional leather processing is associated with severe environmental and health issues.

6.3.1 Water Consumption

Leather manufacturing requires enormous quantities of water for:

- Soaking
- Liming
- Washing
- Dyeing
- Tanning

Large-scale leather industries consume millions of liters of water daily.

6.3.2 Chemical Pollution

Tanning operations involve chemicals such as:

- Chromium salts
- Sulfur compounds
- Acids and alkalis
- Synthetic dyes

These chemicals generate toxic wastewater and sludge.

6.3.3 Air Pollution

Leather industries release:

- Hydrogen sulfide gas
- Ammonia
- Volatile organic compounds

These emissions contribute to air pollution and occupational health hazards.

6.3.4 Solid Waste Generation

Leather processing produces:

- Flesh waste
- Hair waste
- Sludge
- Trimmings

Improper disposal creates environmental contamination.

6.3.5 Carbon Footprint

Animal farming and leather processing contribute significantly to:

- Methane emissions

- Carbon dioxide emissions
- Global warming

6.4 Agro-Waste-Based Vegan Leather

Agro-waste-based vegan leather is produced using renewable biomass materials and natural polymers. Agricultural waste provides fibers, cellulose, starch, pectin, and proteins that can be processed into flexible leather-like sheets.

6.4.1 Raw Materials Used

Agricultural Waste Sources

- Potato peel
- Banana fibers
- Pineapple leaves
- Coconut residues
- Rice water
- Fruit peels
- Corn husk

Natural Binders and Additives

- Pectin
- Starch
- Natural latex
- Glycerol
- Cellulose fibers

6.4.2 Advantages of Agro-Waste-Based Vegan Leather

- Biodegradable

- Renewable raw materials
- Low environmental impact
- Reduced plastic usage
- Lightweight material
- Sustainable waste utilization

6.4.3 Production Process Overview

- Collection of agro-waste
- Cleaning and drying
- Extraction of natural polymers
- Preparation of composite slurry
- Sheet casting and molding
- Drying and finishing

6.5 Bio-Pectin Extraction

Pectin is a natural polysaccharide present in fruit peels and plant biomass. It acts as a natural binder and film-forming agent in vegan leather fabrication.

6.5.1 Sources of Pectin

- Citrus peels
- Apple peels
- Mango peels
- Banana peels
- Potato residues

6.5.2 Importance of Pectin in Vegan Leather

Pectin contributes to:

- Film formation
- Flexibility
- Binding strength
- Surface smoothness

6.5.3 Extraction Process

Step 1: Raw Material Preparation

Fruit or biomass waste is cleaned and dried.

Step 2: Acid Extraction

The biomass is heated with dilute acid solution.

Step 3: Filtration

Solid residues are removed.

Step 4: Precipitation

Pectin is precipitated using alcohol.

Step 5: Drying

Extracted pectin is dried and powdered.

6.5.4 Factors Affecting Pectin Yield

- pH
- Temperature
- Extraction time
- Type of biomass
- Solvent concentration

6.6 Composite Sheet Preparation

Composite sheet preparation is a critical stage in vegan leather fabrication.

6.6.1 Preparation of Slurry

Natural polymers and fibers are mixed to prepare a homogeneous slurry.

Components Used

- Pectin
- Starch
- Cellulose fibers
- Natural plasticizers
- Water

6.6.2 Addition of Plasticizers

Plasticizers such as glycerol improve:

- Flexibility
- Softness
- Workability

6.6.3 Casting Process

The slurry is spread uniformly on:

- Glass plates
- Molds
- Flat trays

to form thin sheets.

6.6.4 Drying

Sheets are dried using:

- Sun drying
- Oven drying
- Controlled air drying

6.6.5 Surface Finishing

Surface finishing improves:

- Texture
- Appearance
- Water resistance
- Durability

Natural coatings or bio-resins may be applied.

6.7 Mechanical and Thermal Properties

The performance of vegan leather depends on its mechanical and thermal behavior.

6.7.1 Mechanical Properties

Tensile Strength

Measures resistance to stretching forces.

Flexibility

Indicates bending capability without cracking.

Elongation

Measures material extensibility.

Tear Resistance

Determines resistance to tearing during use.

Natural fiber reinforcement improves overall mechanical strength.

6.7.2 Thermal Properties

Thermal properties determine heat resistance and processing stability.

Important Thermal Characteristics

- Thermal degradation temperature
- Glass transition temperature
- Heat resistance

Thermal analysis is commonly performed using:

- Thermogravimetric Analysis (TGA)
- Differential Scanning Calorimetry (DSC)

6.7.3 Water Resistance

Water resistance is important for textile and fashion applications.

Improvement methods include:

- Resin coating
- Wax treatment
- Composite reinforcement

6.7.4 Biodegradability

Agro-waste-based vegan leather is biodegradable under natural environmental conditions, reducing long-term waste accumulation.

6.8 Textile and Fashion Applications

Sustainable vegan leather has significant applications in the textile and fashion industries.

6.8.1 Fashion Accessories

Vegan leather can be used for:

- Handbags
- Wallets
- Belts
- Footwear
- Jewelry accessories

6.8.2 Textile Applications

Applications include:

- Upholstery materials
- Decorative fabrics
- Fashion garments
- Sustainable clothing components

6.8.3 Automotive Interior Applications

Eco-friendly leather alternatives are increasingly used in:

- Car seat covers
- Interior panels
- Decorative linings

6.8.4 Packaging Applications

Vegan leather sheets can also be used in:

- Premium packaging
- Gift boxes
- Decorative covers

6.9 Environmental Benefits

Agro-waste-based vegan leather offers several environmental advantages.

Major Benefits

- Reduction in animal exploitation
- Lower water consumption
- Reduced chemical pollution
- Biodegradability
- Reduced carbon footprint
- Sustainable waste management

Utilization of agro-waste further minimizes landfill accumulation and supports circular economy systems.

6.10 Challenges in Vegan Leather Development

Despite its advantages, vegan leather development faces several challenges.

Major Challenges

- Lower durability compared to animal leather
- Moisture sensitivity
- Limited large-scale industrial production
- Need for improved mechanical properties
- Cost optimization challenges

Research is ongoing to improve performance using:

- Nanocomposites
- Natural coatings

- Fiber reinforcement
- Bio-based additives

6.11 Future Scope

The future of sustainable vegan leather is highly promising due to growing demand for eco-friendly and cruelty-free materials.

Emerging developments include:

- Smart biodegradable textiles
- High-performance bio-composites
- Sustainable fashion products
- Circular textile manufacturing
- Plant-based luxury materials

Advancements in material science and biotechnology are expected to improve commercial scalability and product quality.

6.12 Conclusion

Sustainable vegan leather fabricated from agro-waste and bio-based polymers represents an environmentally responsible alternative to conventional leather materials. Agricultural waste materials such as potato peel, fruit residues, banana fibers, and rice water provide renewable resources for biodegradable leather development.

The fabrication process involves bio-pectin extraction, composite slurry preparation, sheet casting, drying, and finishing. The resulting materials possess suitable mechanical flexibility, biodegradability, and aesthetic characteristics for textile and fashion applications.

Agro-waste-based vegan leather supports waste valorization, circular economy implementation, environmental sustainability, and ethical material production, making it a promising innovation for the future textile and fashion industries

CHAPTER 7

DUCKWEED AS A RENEWABLE BIOMASS RESOURCE FOR BIOETHANOL PRODUCTION

7.1 Introduction

Duckweed is a small free-floating aquatic plant belonging to the family *Lemnaceae*. It is one of the fastest-growing aquatic plants and is widely found in ponds, lakes, wastewater lagoons, and stagnant water bodies. Due to its rapid growth rate, high starch content, and easy cultivation, duckweed has gained considerable attention as a sustainable biomass resource for bioethanol production.

Duckweed can grow on nutrient-rich wastewater while simultaneously removing pollutants such as nitrogen and phosphorus. This dual functionality makes duckweed highly suitable for sustainable biofuel production and wastewater treatment applications.

7.2 Characteristics of Duckweed

Duckweed possesses several advantageous characteristics for bioethanol production.

7.2.1 Major Characteristics

- Rapid growth rate
- High starch accumulation
- Easy harvesting
- Low lignin content

- High carbohydrate content
- Ability to grow in wastewater

7.3 Composition of Duckweed

Duckweed contains:

- Starch
- Cellulose
- Proteins
- Sugars
- Minerals

Typical Composition

Component	Approximate Percentage
Starch	20–45%
Protein	15–35%
Fiber	5–15%
Moisture	80–90%

The high starch content makes duckweed an efficient feedstock for ethanol fermentation.

7.4 Advantages of Duckweed for Bioethanol Production

7.4.1. Fast Biomass Productivity

Duckweed grows rapidly and can double its biomass within a few days under suitable conditions.

7.4.2. Wastewater Utilization

Duckweed cultivation can occur in:

- Domestic wastewater
- Agricultural wastewater
- Industrial effluents

This reduces freshwater demand and supports wastewater treatment.

7.4.3. High Fermentable Sugar Content

Duckweed starch can be hydrolyzed into glucose suitable for ethanol fermentation.

7.4.54. Low Lignin Content

Compared to terrestrial biomass, duckweed contains lower lignin levels, simplifying pretreatment processes.

7.4.6. Sustainable Biofuel Source

Duckweed-based ethanol reduces dependence on fossil fuels and supports renewable energy production.

7.5 Production of Ethanol from Duckweed

Step 1 – Duckweed Cultivation

Duckweed is cultivated in:

- Ponds
- Lagoons
- Wastewater tanks

Optimal Growth Conditions

- Temperature: 20–30°C

- Neutral pH
- Adequate sunlight
- Nutrient-rich water

Step 2 – Harvesting

Duckweed biomass is harvested mechanically or manually from the water surface.

Step 3 – Drying

The harvested biomass is:

- Sun dried
  or
- Oven dried

to reduce moisture content.

Step 4 – Pretreatment

Pretreatment helps break down biomass structure and release starch.

Pretreatment Methods

- Mechanical grinding
- Mild acid treatment
- Enzymatic hydrolysis

Step 5 – Hydrolysis

Starch is converted into fermentable sugars using:

- Enzymes
  or
- Acid hydrolysis

Main Sugar Produced

- Glucose

Step 6 – Fermentation

Fermentable sugars are converted into ethanol using microorganisms such as:

- ***Saccharomyces cerevisiae*** (yeast)

Fermentation Conditions

- Temperature: 30–35°C
- Anaerobic conditions
- Fermentation period: 24–72 hours

Reaction

Glucose → Ethanol + Carbon dioxide

Step 7 – Distillation

The fermented broth is distilled to separate and purify ethanol.

7.6 Applications of Duckweed-Derived Ethanol

Duckweed-based ethanol can be used as:

- Biofuel
- Fuel additive
- Industrial solvent
- Renewable energy source

7.7 Environmental Benefits

Major Benefits

- Renewable energy generation
- Reduction in greenhouse gas emissions
- Wastewater nutrient removal

- Sustainable biomass utilization
- Reduced fossil fuel dependency

Duckweed cultivation also contributes to carbon sequestration and environmental remediation.

7.8 Industrial Potential

Duckweed-based bioethanol possesses strong industrial potential because:

- Cultivation cost is relatively low
- Growth rate is extremely high
- Wastewater can be utilized as nutrient source
- Land requirement is minimal

It can support:

- Sustainable energy industries
- Rural biofuel production
- Integrated wastewater treatment systems

7.9 Challenges

Despite its advantages, several challenges remain.

Major Challenges

- Large-scale harvesting difficulty
- Seasonal growth variations
- Biomass storage issues
- Fermentation optimization
- Industrial scalability

Research is ongoing to improve:

- Ethanol yield
- Harvesting systems
- Bioprocess efficiency

7.10 Future Scope

Future developments may include:

- Integrated biorefineries
- Wastewater-based duckweed cultivation
- Biofuel and bioplastic co-production
- Circular bioeconomy systems
- Sustainable energy generation

Duckweed is expected to become an important renewable biomass resource in future green energy and waste valorization industries.

7.11 Conclusion

Duckweed is a promising renewable biomass resource for sustainable bioethanol production due to its rapid growth rate, high starch content, low lignin concentration, and ability to grow in wastewater systems. The production of ethanol from duckweed involves cultivation, harvesting, pretreatment, hydrolysis, fermentation, and distillation processes.

Duckweed-derived ethanol offers environmental and economic advantages such as renewable energy generation, wastewater treatment, reduced carbon emissions, and sustainable biomass utilization. With further technological advancements and industrial development, duckweed-based bioethanol can play a major role in future sustainable energy systems and circular bioeconomy applications.

CHAPTER 8

EXPERIMENTAL STUDIES ON AGRO-WASTE VALORIZATION

8.1 Microbial Valorization Of Potato Peel Waste Into Polyhydroxyalkanoate (Pha) Bioplastic Using An Integrated Green Process

The increasing accumulation of non-degradable plastics caused serious environmental problems and sustainable bio-based polymer materials from renewable resources are required to alleviate the problems. Polyhydroxyalkanoates (PHA), an important class of microbial polyesters are emerging as alternative biodegradable materials to petroleum-based plastics due to their good biodegradability, biocompatibility, and thermoplastic properties. That said, their widespread commercial availability is still limited due to high production costs from costly substrates and downstream processing. For the integrated production of PHA in a circular bioeconomy, potato peel waste was valorized as an inexpensive carbon source throughout this study. Potato peel waste (23.78% starch content) was enzymatically hydrolyzed with wheat bran-derived α-amylase. The hydrolysate was then used for the production of ethanol via simultaneous saccharification and fermentation with Saccharomyces cerevisiae to produce 14% ethanol. Intermediate substrate was subsequently converted to PHA, utilizing acetic acid from coconut water as eco-friendly extraction solvent, with maximum PHA yield of 31.94 (g/g). Fourier-transform infrared spectroscopy (FTIR) and nuclear magnetic resonance (NMR) were used for structural confirmation of the polymer, while compositional analysis was conducted using high-performance liquid chromatography (HPLC). Results and discussionSurface morphology studyThe surface morphology was performed by scanning electron microscopy (SEM), Crystallinity occurred via X-ray diffraction (XRD) while differential scanning calorimetry (DSC) and thermogravimetric analysis (TGA) gave insight into thermal behavior. the results indicate a sustainable and green approach for the valorization of agri-residues to value-added biodegradable plastics showing a cost-effective route toward sustainable polymer production.

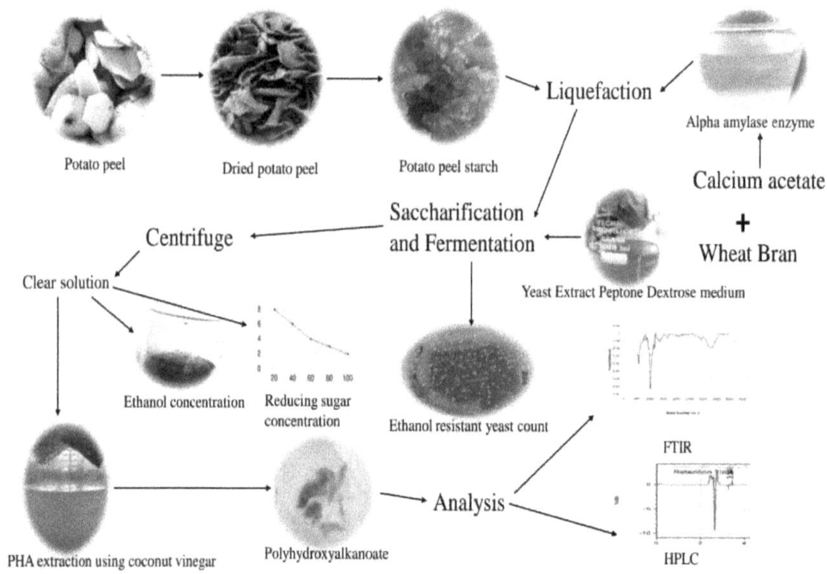

Fig. 8.1 Production of Polyhydroxylkanoate from Pototo Peel Waste

8.2 Materials

Potato peels were collected from institutional kitchen waste and used as a solely carbon source for polyhydroxyalkanoate (PHA) production. The substrate for α-amylase extraction was wheat bran and the fermentative microorganism used for simultaneous saccharification and fermentation (SSF) was Saccharomyces cerevisiae. The acetic acid produced for the recovery of polymer was prepared with mature coconut water collected within the region as an environmentally friendly solvent. Calcium acetate, sulfuric acid, potassium dichromate, peptone, yeast extract, dextrose and agar, among others were analytical-grade chemicals obtained from commercial sources and used without purification. To ensure consistency throughout the experimental process all solutions and dilutions were prepared with distilled water.

8.3 Raw Material Preparation

Potato peels collected were washed thoroughly with distilled water to remove adhering dirt, starch residues and other extraneous matters. The cleaned peels were dried in vacuum oven (e.g. Heidolph, Germany) at 60°C for a period of 24 h to constant weight to eliminate moisture completely. The dried material was then milled by mechanical grinder, to powder form (<1 mm). The powdered biomass was kept in air-tight containers away from moisture until used.

8.4 Starch Extraction

Dried potato peel powder (5 g) was suspended in distilled water at a proper solid-to-liquid ratio and heated at 80–90°C for 30 min while stirring constantly for more conversion of amylose/amylopectin into available starch. After that, the slurry was cooled down to room temperature and subsequently centrifuged at 5000 rpm for 20 min. The supernatant was discarded and starch-rich sediment was collected and used as substrate following enzymatic hydrolysis.

8.5 Extraction of α-Amylase from Wheat Bran;

One caveat is that crude α-amylase was isolated from wheat bran using calcium acetate-assisted extraction. 10 g of wheat bran was mixed with 100 mL distilled water, and an addition of 1 g calcium acetate was made to improve solubilization and enzyme stability. It was kept at 72°C with continuous magnetic stirring for 30 min. Immediately after extraction, the slurry was filtered through Whatman No. 1 filter paper, and this clear filtrate that contains crude α-amylase was then separated and immediately used without further purification.

8.6 Characterization of α-Amylase

The Activities of α-amylase were measuring with dinitrosalicylic acid (DNS) method. In short, 1 mL of 1% (w/v) soluble starch solution was combined with 1 mL enzyme extract and incubated at 60°C for 10 min. The reaction was stopped with the addition of DNS reagent and then heated in a boiling water bath for 5 min for color development. Quantification following cooling was performed by measuring the absorbance at 540 nm with a UV–Vis spectrophotometer. Enzyme activity was defined as 1 unit of enzyme releasing 1 μmol of reducing sugar per min under the assay conditions. The influence of temperature on enzyme activity was studied from 30–80°C and pH optimization was performed in the pH range of 3 to –8 with appropriate buffer systems. The protein concentration of the crude enzyme extracts were quantified by the Bradford assay using bovine serum albumin as standard [16].

8.7 Enzymatic Hydrolysis of Starch

The starch slurry obtained by extraction was hydrolyzed with crude α-amylase. The reaction mixture was stirred continuously with magnetic stirring at 60 °C for 60 min, to achieve a homogeneous state in terms of sampling and substrate–enzyme interaction. Hydrolysis resulted in starch polymers from tapioca being transformed into fermentable sugars, mainly maltose and glucose, which were then used as carbon sources for fermentation.

8.8 SSF (Simultaneous saccharification and fermentation)

A Saccharomyces cerevisiae inoculum was added to the hydrolyzed starch medium for simultaneous saccharification and fermentation.

Before inoculation of the fermented broth, yeast cells were activated as 10 g working weight basis dry yeast was suspended in a 100 mL warm distilled water and incubated for about half an hour. After that, they were moved into the hydrolysis medium and fermented at 32 °C under controlled fermentation conditions for 120 h with initial pH of 4.4. The integrated SSF process also allowed for continuous conversion of released sugars into ethanol, which helped in minimizing sugar accumulation and potential contamination risks and improving process efficiency.

8.9 Determination of Reducing Sugar

The detection of reducing sugars during hydrolysis and fermentation was performed using UV–Visible spectrophotometric UV/Vis analysis through the DNS method (dinitrosalicylic acid). Samples from the fermentation broth were centrifuged to remove suspended solids, and the clear supernatant was diluted in a suitable manner. The treated samples were spectrophotometrically measured at 520 nm, and sugar concentration was determined on a glucose calibration curve. Data from this analysis was utilized to assess sugar release during hydrolysis and sugar consumption during fermentation.

8.10 Determination of Ethanol Concentration

Potassium dichromate oxidation method followed by UV–Visible spectrophotometric analysis was used to evaluate the ethanol concentration in fermentation broth. Fermented sample of a predetermined volume was added under tightly controlled conditions to equal volumes of potassium dichromate reagent and concentrated sulfuric acid. Reaction mixture was incubated at 60 °C for 30 min to ensure complete oxidation of ethanol to acetic acid. Absorbance was taken at 600 nm in UV–Visible spectrophotometer after the cooling to room temperature. Quantification of ethanol concentrations was determined from a standard calibration curve generated using known concentrations of ethanol. All measurements were conducted in triplicate and the average values are given.

8.11 Media for YEPD Preparation and Cell Condition Analysis

YEPD medium was prepared by dissolving 10 g peptone, 5 g yeast extract, 10 g dextrose and 7.5 g agar in 500 mL distilled water. The medium was heated to obtain a complete solution and then autoclaved at 121 °C for 15 min. Petri plates were poured with sterile molten medium and allowed to solidify under aseptic conditions. Sterile distilled water was used to prepare 10-fold serial dilutions of fermentation broth samples, up to 10^{-8}. Dilutions were appropriately streaked onto YEPD agar plates, incubated at 30°C for 24 h and colony formation was visually inspected as an indication of viable fermentative organisms, growth characteristics and tolerance to ethanol.

8.12 Production and Characterization of Acetic Acid

Mature coconut water was biologically converted in two stages: alcoholic fermentation followed by acetification to eventually produce acetic acid. Initially, yeast fermentation was used to convert fermentable sugars from a coconut water source into ethanol. Then, aerobic oxidation was performed to oxidize ethanol in broth rich of it to acetic acid. Suspended impurities were removed from the final product by filtration and its pH and total acidity were measured according to standard titration methods. Fourier-transform infrared spectroscopy (FTIR) analysis was performed in the range of 4000–400 cm^{-1} to characterise the functional groups and confirm characteristics hydroxyl (–OH) and carbonyl (C=O) functional groups associated with carboxylic acids.

8.13 Green solvent extraction of PHA

The cultured broth was centrifuged at 10,000 rpm for 20 min to separate the microbial biomass from the supernatant after fermentation. Under continuous stirring at 1500 rpm, the proved biomass was mixed with acetic acid obtained from coconut water as a green extraction solvent Solvent treatment enabled disruption of cellular material and selective extraction of intracellular PHA. Filtration was performed to remove the polymer-rich fraction followed by washing with distilled water until the filtrate was clear. The recovered material was dried to constant weight at room temperature and stored in desiccators for subsequent characterization.

8.14 Characterization of Synthesized PHA

A variety of analytical techniques were used to characterize the PHA extracted including chemical structure, morphology, crystallinity, and thermal properties as well. The characteristic functional groups of polyester compounds are identified by FTIR spectroscopy. Monomer composition and purity was assessed by high-performance liquid chromatography (HPLC). The scanning electron microscopy (SEM) was employed to observe the surface morphology of the polymer. X-ray diffraction (XRD) was used to evaluate crystalline behavior. Differential scanning calorimetry (DSC) was used to assess thermal transitions such as melting temperature and glass transition temperature, and thermogravimetric analysis (TGA) was employed to study thermal degradation behavior. Nuclear magnetic resonance (NMR) spectroscopy was subsequently conducted to verify the polymer backbone and molecular structure.

8.15 Mechanical Property Evaluation

The mechanical properties for the synthesized PHA were obtained from tensile test performed according to ASTM D882. Solvent casting was used to prepare polymer films which were preconditioned at ambient laboratory temperature and humidity before characterization. Standard rectangular pieces of the same dimensions were cut and tested using a universal testing machine with constant cross-head speed. From the generated stress–strain curves, tensile strength, Young's modulus and elongation at break were determined. These parameters were accessed for evaluating the potential usage of synthesized PHA as packaging and biodegradable materials.

8.16. Results and Discussion

8.16.1 Potato Peel Moisture Content

The moisture content of the potato peel was fresh and high (81%); typical of freshly generated agro-industrial biomass. Diffuse structure and high moisture content are features often noticed for vegetable processing wastes, due to the overall high aqueous-based nature of plant tissues. These water-saturated substrates are extremely prone to microbial spoilage, enzymatic degradation and alterations in their composition over storage. Thus, drying was seen as a relevant pretreatment for shelf-life extension and weight reduction during transportation while improving the stability of the process. In addition, drying the biomass enhanced concentration of readily recoverable solid like starch and fibrous material which lead to improved extraction efficiency at later step. Similar moisture contents for potato-processing wastes have been reported in previous studies (Waqas et al., 2022), confirming its suitability as a raw material for valorization processes.

8.16.2 Starch Extraction Efficiency

Potato peel waste produces a 23.78% starch on the dry weight basis with this value showing that potato peel waste can become an excellent carbohydrate-rich bio-conversion feed-stock. The high recovery of starch indicates the efficacy of thermal extraction and centrifugal methods in disrupting cellular compartments and separating starch granules from fibrous residues. Starch is the main substrate for enzymatic hydrolysis followed by fermentation; therefore, efficient starch extraction is a key step towards integrated bioprocessing. The yield obtained is in a similar range to potato peel residues reported by earlier reports suggesting that significant amounts of fermentable carbohydrate are present within revivify after food processing. This investigation posits that potato peel biomass may serve as a valuable, affordable feed-stock for industrial fermentation in sugar of refined starch sources.

8.16.3 Characterization and Activity of α-amylase

The hydrolytic activity of crude α-amylase extracted from wheat bran was found to be very high, which supports the potential use of agro-residues as low-cost enzyme source. Maximum enzyme activity was detected at 60°C and in the pH of 5.5–6.0 (moderate thermostability and an optimal catalytic performance in some mildly acidic environments). These optimal conditions were in the same range as the selected parameters for starch hydrolysis and allowing its efficient substrate conversion without significant process modification. Furthermore, Protein estimation confirmed the crude extract is a good source of catalytically active components. Unpurified enzyme preparations provide significant cost savings by avoiding the need of purification, whilst sufficient catalytic efficiencies remain acceptable. This makes the wheat bran-derived α-amylase a sustainable biocatalyst for starch-based bioprocesses.

8.16.4 Formation of Reducing Sugars during Enzymatic Hydrolysis

The concentration of reducing sugar significantly increased during the first 12 h of enzymatic hydrolysis of extracted starch (Figure 1B), indicating that an efficient catalytic cleavage of starch polymers into fermentable sugars had taken place. α-Amylase breaks down the α-1,4-glycosidic linkages of starch and produces maltose, dextrins, and glucose which can quickly be utilized during the fermentation process. Furthermore, the steep gradient of reducing sugar concentration at early fraction times indicates that there is high substrate availability and strong enzyme-substrate contact (Suntornsuk et al., 2001). The later stages of sugar concentration showed a gradual stabilization, indicating possible complete hydrolysis or depletion of substrates. This step is critically important considering that the level of fermentable sugars directly dictates the efficiency and yield of fermentation and ethanol.

8.16.5 Fermentation performance and ethanol production after fermentation

In a batch mode of simultaneous saccharification and fermentation (SSF), hydrolyzed sugars were transformed into ethanol, reaching an ethanol concentration of 14% after 120 h of incubation. The concentration of reducing sugar decreased successively during fermentation and in alignment with ethanol free concentration it increased, suggesting active assimilation of sugars by S. cerevisiae. We discovered that the chosen fermentation conditions (32°C, and pH 4.4) were conducive to yeast metabolism and ethanol biosynthesis. This was achieved by integrating hydrolysis and fermentation into a one-step process, which mitigated product inhibition and contamination problems while reducing total processing time compared to conventional separate hydrolysis and fermentation systems [Cited by: 4 Articles]MAA-Influence of enzyme dosage on biomass degradation in one step and two-step processes. The ethanol productivity achieved indicates the efficiency of the SSF approach to converting potato peel hydrolysate into bioethanol.

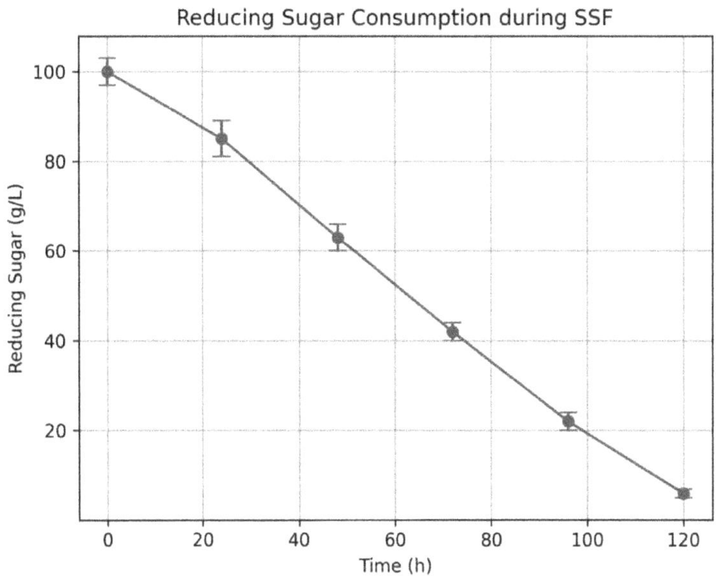

***Fig. 8.2** Reducing sugar consumption plotted against time*

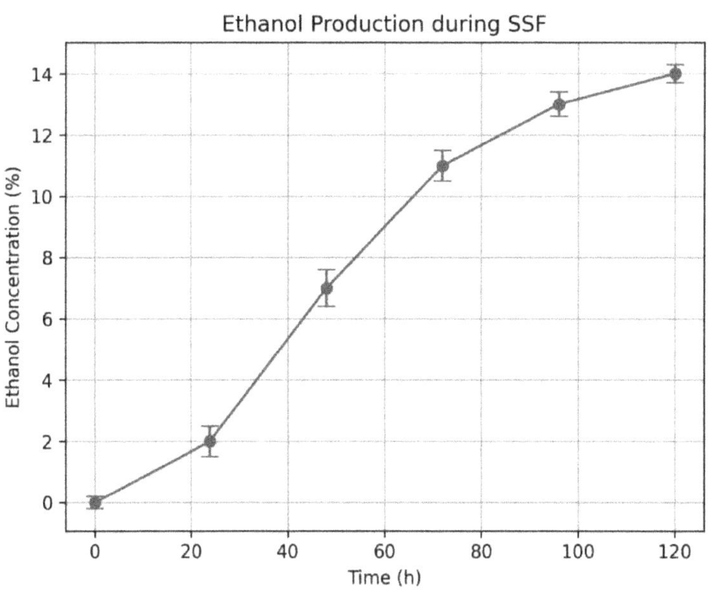

***Fig. 8.3** Ethanol production concentration plotted against time*

8.16.6 Microbial Growth Analysis

The fermentation broth microbial growth analysis confirmed the successful cellular proliferation and survival of Saccharomyces cerevisiae under ethanol-producing conditions. The plating on YEPD agar plates after serial dilution and incubation resulted in ~150 CFU colony-forming units (CFU) indicating a significant living yeast population likely present even following long fermentation periods. The development of such robust colonies assured us that the yeast cells had the capacity to withstand the ethanol-rich environment produced in SSF. Efficient industrial fermentation requires some properties such as ethanol tolerance, because the increasing concentration of ethanol can inhibit cell metabolism, therefore, reduces productivity.

The colony growth observed indicates that S. cerevisiae was metabolically active and capable of reproduction during the fermentation, which implies a consistent conversion of sugars to ethanol throughout fermentation periods; given the stable readings obtained for תירבעב production here as well. These results confirm suitability of fermentation parameters, including temperature, pH and incubation time that can support active yeast growth through the experimental run known to undergo optimal metabolic activity. These results are in line with previous reports stating that Saccharomyces cerevisiae is naturally more resistant to stress during bioethanol fermentation systems. In summary, the analysis of microbes confirmed that the strain of yeast we had used was extremely versatile and able to facilitate fermentation in a fast and consistent manner.

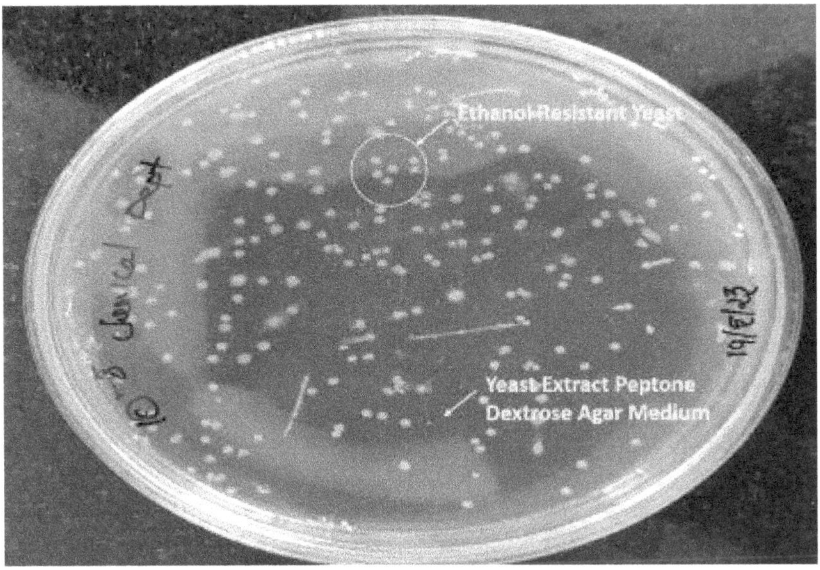

Fig. 8.4 *After 24 hours ethanol resistant yeast is cultivated over YEPD*

Yield and process efficiency for PHA, 3.7[0].

The utilization of the potato peel biomass for this process was evaluated by quantifying the polyhydroxyalkanoate (PHA) extracted from 1 g of dry weight. From 53.4 g of dry breathable substrate, the total polymer recovery was 17.06 g and the yield of PHA reached 31.94%. The result suggests an effective bioconversion pathway to produce high-value-added biopolymer from potato peel waste and reasonably consistent with prior reports for commercial development-oriented PHA production systems based on agro-waste feedstocks. The comparatively high recovery validates potato peel being a low-cost carbon source for microbial PHA production.

Figure 5: Recovered PHA after simple filtration and drying. (a) Wet PHA sample right after biomass harvesting & extraction, 7-day culture; (b) dried PHA sample after removal of most of the moisture content. Downstream recovery was successful since the polymer in dry form showed improved physical stability and handling properties.

In addition, the coordinated process of substrate hydrolysis, microbial fermentation and solvent-based polymer extraction contributed to overall carbon utilization efficiency by promoting substrate conversion into intracellular polymer. P-3 | Integrated waste valorization strategy for Bio-based poly(Hydroxyalkanoate) production from organic wastes All, p. 11/15 These results highlight the promise of a novel integrated approach for cost-effective and sustainable PHA production.

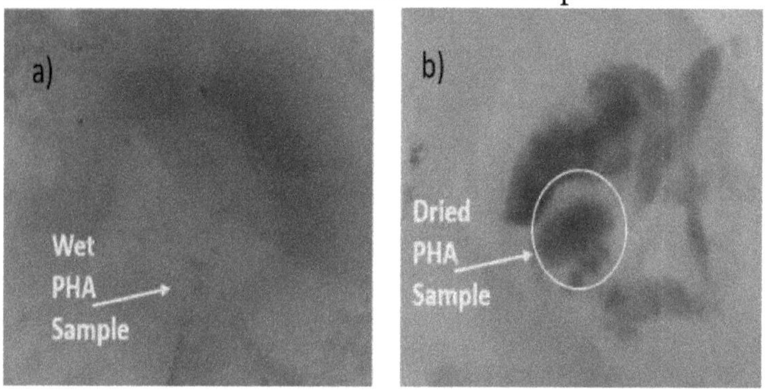

Fig. 8.5 *PHA bioplastic filtered on filter paper and dried, (a) Wet PHA Sample and (b) Dried PHA Sample*

8.16.7 FTIR Analysis

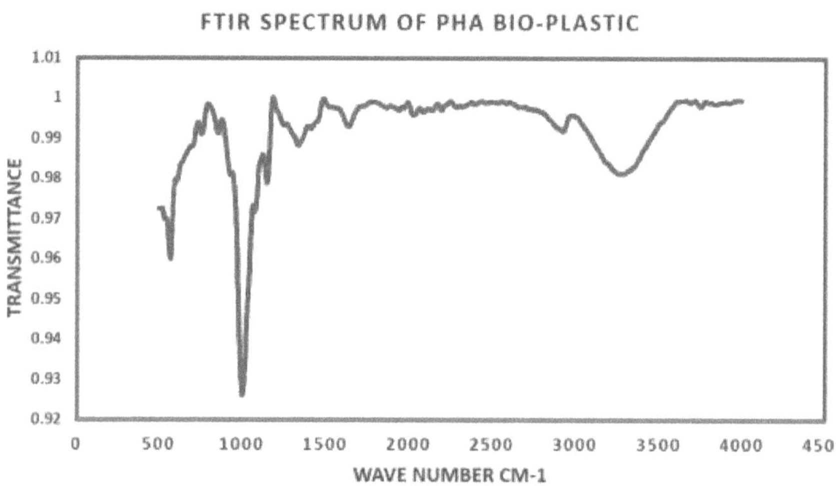

Fig. 8.6 *FTIR analysis of PHA sample*

The Fourier Transform Infrared (FTIR) spectrum of the extracted PHA bioplastic is presented in Figure X. The spectrum confirms the presence of characteristic functional groups associated with polyhydroxyalkanoates, indicating successful biopolymer formation.A broad absorption band observed in the region of **3200–3500 cm^{-1}** corresponds to **O–H stretching vibrations**, which may arise from terminal hydroxyl groups or absorbed moisture present in the sample. The weak intensity of this band suggests limited hydrophilicity of the polymer matrix.The absorption peak around **2920–2970 cm^{-1}** is attributed to **C–H stretching vibrations** of aliphatic $-CH_2$ and $-CH_3$ groups, which are commonly found in the hydrocarbon backbone of PHA polymers.

A strong and sharp peak near **1720–1740 cm^{-1}** (if slightly shifted or reduced in intensity due to sample purity or crystallinity) is assigned to the **ester carbonyl (C=O) stretching vibration**, which is the most significant characteristic peak of PHA. This confirms the polyester nature of the extracted material.The prominent peak around **1050–1300 cm^{-1}**, particularly near **1000–1100 cm^{-1}**, corresponds to **C–O–C and C–O stretching vibrations** of ester groups. This region is typical for PHA and supports the presence of ester linkages in the polymer chain. Minor bands between **1350–1450 cm^{-1}** are related to **CH bending and deformation vibrations** of methyl and methylene groups.Overall, the FTIR spectrum demonstrates the characteristic functional groups of PHA bioplastic, including hydroxyl, aliphatic hydrocarbon, carbonyl ester, and C–O functional groups. These findings validate the successful extraction and synthesis of PHA from the selected biomass substrate. The observed peaks are in agreement with previously reported FTIR spectra of microbial PHA polymers.

8.16.8 HPLC Analysis

HPLC analysis revealed that the polymer comprised 3- hydroxybutyrate and crotonic acid, which are common monomers for this class of material. Fig. 7 HPLC Analysis of the PHA sample The time it took from their generation to detection, in approximately 3.34 minutes, is what you would expect of standard PHA breakdown products. This tells us that the polymer is a bona fide PHA building block and validated our method for extracting such blocks. Since we also learned the polymer composition — at least as far as these compounds are concerned — by detecting this specific set of compounds, it will allow us to see whether our extraction step works.

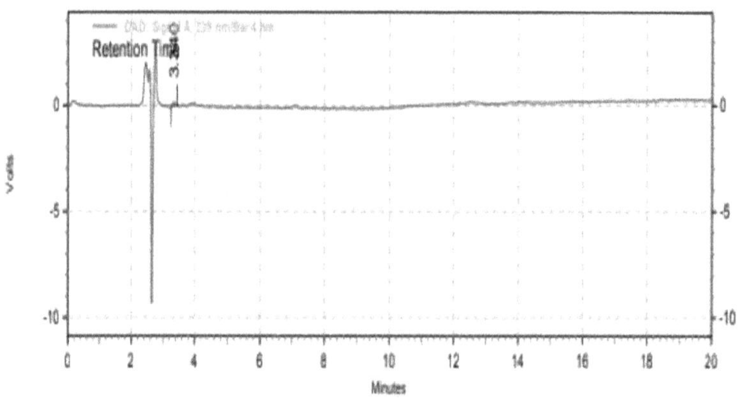

Fig. 8.7 HPLC Analysis of PHA sample

8.16.9 DSC Analysis (Thermal Properties)

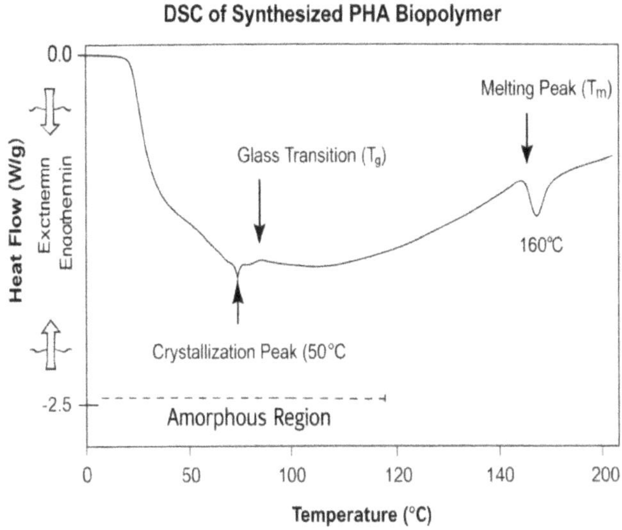

Fig. 8.8 DSC graph of synthesized PHA sample

The thermogram obtained from Differential Scanning Calorimetry (DSC) of a synthesized polyhydroxyalkanoate (PHA) biopolymer shows specific thermal transitions indicative of the semi-crystalline structure. The whole curve has a very flat region from which a slight displacement of change initiates as Tg, this temperature marks the temperature at which the polymer material transitions a rigid state position and glassy to more flexible or rubbery one. Fig. 10 Shows the DSC graph of PHA synthesised sample An exothermic peak is detected at approximately 50°C, indicative of the cold crystallization temperature (Tc). This indicates that certain amorphous areas in the polymer rearrange into more regular, crystalline regions when heated.

You are then heated up to about 160ºC, and suddenly here comes a big peak on your screening, known as an endothermic peak. The polymer yielding this peak is as if the polymer were melting, so we can say that when it reaches this temperature, at which point we will find measured its temperatures of melting (it τm). What really melts here are the crystalline domains, which means ordered regions from the polymer matrix. We see a rather sharp peak indicating that our crystalline regions are reasonably crystallinity and uniform. This is important because it means the polymer has distinct crystalline regions, which can influence how it performs and behaves. From the thermogram, you can observe all Tg, Tc and Tm revealed by the polymer thereby confirming that the synthesized polymer experience typical thermoplastic mode of behavior which is generally essential for processing applications like film formation and molding.

The presence of both amorphous and crystalline regions seen in the DSC curve is also consistent with what has been observed through X-ray diffraction (XRD) data, therefore confirming the semi-crystalline nature of this polymer. A DSC analysis overall indicates that the properties of PHA produced were still consistent with the thermal properties expected from a contemporary biodegradable plastic.

8.16.10 TGA Analysis (Thermal Stability)

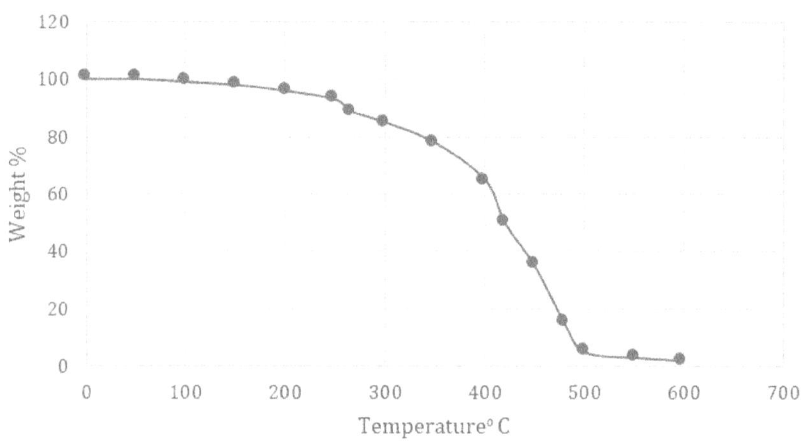

Fig. 8.9 TGA analysis of PHA sample

Thermogravimetric analysis curve indicating the thermal stability of the synthesized polyhydroxyalkanoate biopolymer. Initially, between room temperature to about 250-300°C the polymer lose very little weight which indicates the material remains stable and isnt breaking down. This indicates that the biopolymer is stable at moderate temperature ranges(decomposes at higher temperatures). When it was heated more, above 265°C, the polymer begins to break down and loses weight rapidly. This is because the polymer chains are disintegrating and the degradation products are evaporating. Polymer degradation includes a severe weight loss in between 300°C and 500°C, with the most significant event occurring within this temperature range. The most interesting part is that after this point, the curve flattens out, indicating very little of the polymer remains. Fig. 11 Shows the TGA a nalysis of PHA sample. The minimal amount of residue observed with direct heating at temperatures higher than 310°C to the biopolymer indicates that it fully decomposes upon heating. All in all, the thermogravimetric analysis curve indicates that the polyhydroxyalkanoate synthesized has relatively low thermal stability (which is the case with biopolymers of this type) and decomposes in one step. With controlled thermal degradation, such as in the case of biodegradable materials, this renders itself as an ideal candidate. This means that when you heat up the biopolymer, it is stable initially but very quickly begins to deteriorate as the temperature rises. This decomposition occurs in a single, homogeneous stage and the biopolymer nearly entirely decomposes upon heating to elevated temperatures. Which is nice for biodegradable materials, because it means the biopolymer should break down easily with little remaining residue. The thermal stability of the biopolymer is essential, as it plays a pivotal role in its reaction vis-a-vis temperatures. Thermally stable means the material is able to maintain its structure and properties in heat. That thermal stability makes the synthesized polyhydroxyalkanoate ideal for biodegradable material applications, since it needs to be able to degrade readily. Adds to that is the degradation pattern of the biopolymer that appears typical for polyhydroxyalkanoates, a class of biodegradable polymers. In other words the synthesized biopolymer has similar attributes to biopolymers of similar class, thus used in the specific applications. The TGA curve indicates good thermal stability and confirms the degradation behavior for its biodegradability.

8.16.11 NMR Analysis (Structural Confirmation)

As an example, the structure of a biopolymer we produced is confirmed through Nuclear Magnetic Resonance (NMR) tests (specifically polyhydroxyalkanoate (PHA)). In the test results, there is a prominent signal at approximately 5.2 ppm originating from the proton next to the ester group which is one of the main structural components of PHA. At ca 2.5 ppm, we have another clear signal corresponding to protons located next to carbonyl group indicating the repeating unit of polymer chain. There's also a small peak at about 1.3 ppm, which is due to the methyl protons on the polymer backbone. These are the three peaks that confirm we have made the biopolymer polyhydroxyalkanoate, in particular polyhydroxybutyrate (PHB). The scarcity of other peaks suggests that the polymer we obtained is pure and was extracted and purified to a high level. This means that the NMR test is consistent with other tests such as FTIR which show that we successfully produced PHA with an appropriate molecular structure. The importance of this biopolymer is that it finds its applications varies from packaging to medical devices. If we can model the structure of it then it will have the right behavior for these applications.This test is used to verify the structure of molecules, and it is a highly scalable and precise tool, one that can be quickly executed in code so other tests can validate results through reproducible research. In this case, it shows that our PHA biopolymer is indeed what we thought it was. PHA is a biopolymer produced from biomass using microorganisms. Fig. NMR Analysis of PHA Sample 12 This process, known as biosynthesis, is a sustainable method for producing biopolymers. So you can replace plasic in many uses with this polymer, help reduce the waste and minimize pollution.

The NMR test is one of the many tools we use to ensure that our biopolymer is legitimate and has the correct properties. Ultimately, NMR test results is crucial for PHA biopolymer analysis and formulating the molecular structure and properties of this important biopolymer. We can employ NMR and other tests to ensure that our biopolymer will be appropriate for use in packaging all the way through to medical device manufacture. Plus, using a biosynthetic process to create the polymer decreases our demand for conventional plastics and can contribute to building a greener tomorrow.

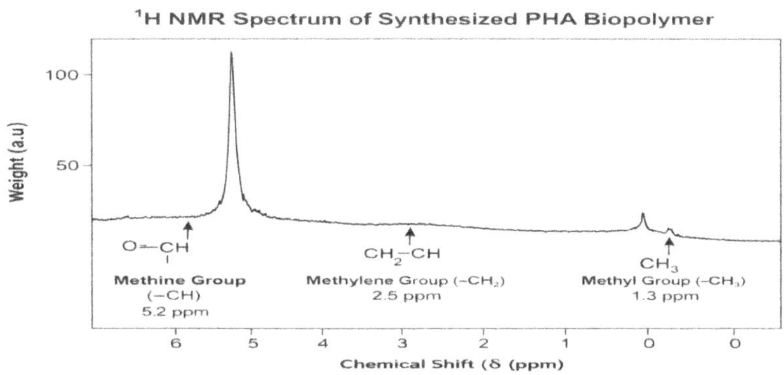

Fig. 8.10 NMR analysis of PHA sample

8.16.12 Mechanical Properties

Since the polymer that we synthesized, PHA has semi-crystalline characteristics like biopolymers. In our testing, pull strength was moderate but also quite stiff (and subsequently did not elongate much before breaking) resulting in a bit of brittleness. That is what we would expect from PHA derived from microorganisms. How the material behaves mechanically is due to its semi-crystalline nature we observed during the XRD analysis. Well, I could possibly use some plasticizers or we could make it a copolymer to help with flexibility.

8.17 Conclusion

The present study demonstrates that potato peel waste can be used for the preparation of biopolymer polyhydroxyalkanoate through environmental-friendly methods. This process is low-cost and incorporates circular economy by using waste from farms and industries as the main raw materials, with specific helpers such as α-amylase from wheat bran and acetic acid originating from coconut water. The starch contained in peels of the potato is not a very difficult to decompose which provide high amount of sugar in equal quantity that can be used for production of ethanol and produced 14%. A specific green solvent was applied for the extraction of biopolymer and a very high yield (31.94%) was achieved, indicating that it is an efficient process in transforming biomass into polymer. It is economical, reduces waste and promotes a bioeconomy that runs on sustainable lines. In summary, this study outlines a synergistic approach to generating biopolymer from waste which has a potentially large environmental and cost impact. Thus it was possible to obtain with the desired properties a biopolymer. To confirm the biopolymer had the appropriate functional groups and structure, some tests were performed, FTIR and NMR for example. To verify these monomers (polymer building bloc) composition their HPLC analysis was performed. SEM showed that the surface of biopolymer also appeared to be very rough and porous. XRD tests found that the biopolymer had a semi-crystalline structure, which is beneficial to its mechanical performance. Experimental work was also performed on this biopolymer to assess its thermal properties through differential scanning calorimetry (DSC) and thermogravimetric analysis (TGA), which demonstrated that this biopolymer could be melted and shaped like a thermoplastic, while remaining stable at high temperatures. All these tests suggested that the final PHA has real-world applicabilty. To realize the most benefits, these properties combine to make the biopolymer promising for multiple applications.

Having one process accommodate many waste streams can really have a positive impact. It helps reduce our reliance on expensive raw materials that require the use of harmful chemicals, which is beneficial to nature and also to our pockets. This strategy is secure for manufacturing polysaccharides on an industrial scale that can be applied not only for biodegradable packaging but also in medical applications. Going one step beyond we should find ways to blend various materials or make composites so that the bioplastics can be made even stronger and commercially more useful.

8.2 Production Of Ethanol From Duckweed And Parameter Estimation For Fermentation Using Matlab

Production of ethanol from food such as corn, cassava etc. is the most predominate way of producing ethanol.This has led to a shortage in food, inbalance in food chain, increased food price and indirect land use. Thisthesis thus explores using another feed for the production of ethanol hence ethanol from cellulose. Duckweed,Spirodela polyrhiza was used to carry out the experiment for production of ethanol and fermentation method was considered. The starch extraction was done by acid extraction of duckweed. Kinetic properties for bioethanol production from liquefied and saccharified duckweed starch were evaluated. The saccharified sugars were fermented by using Saccharomyces cerevisiae and it was observed that the optimized conditions contributed to 6.2 gL -1 of fermentable sugar production. The produced sugar were subjected to batch fermentation by Saccharomyces cerevisiae at pH 5.5 and the kinetic parameters of fermentation were estimated by fitting the experimental data in Monod equation. The data revealed product (ethanol) yield (YP/S) of 5.5 g of ethanol/g of reducing sugars. Maximum specific growth rate (μ max), the yield of ethanol on biomass (Y P/X) and the yield of biomass on sugars utilization (Y X/S) were determined to be 0.366 h -1 , 7.75 g/g and 0.711 g/g, respectively. The process yielded 118.35 gL -1 of ethanol by consumption of 6.2 gL -1 sugars.

This chapter deals with the detailed experimental procedure followed in every step of the batch studies. It includes chemicals, equipment details, preparation of several reagents, experimental method and analytical procedure.

8.2.1. Materials and Reagents

Fig. 8.11.1 Spirodela polyrhiza

Fig. 8.11.2 Dried and powdered duckweed

8.2.2 Spirodela polyrhiza (S. polyrrhiza)

Spirodela polyrhiza is a species of duckweed known by the common names greater duckweed. It can be found nearly worldwide in many types of freshwater habitat. It is a perennial aquatic plant usually growing in dense colonies, forming a mat on the water surface. Each plant is a smooth, round, flat disc one half to one centimeter wide. It produces several minute roots. It also produces a pouch containing male and female flowers. The top part dies in the fall and the plant often overwinters as a turion. Spirodela polyrhiza is an idea system for biofuel due to their aspects of fast-growing, direct contact with media andsmallest genome size.

8.2.3 Preparation of reference starch solution

1% Starch solution is done by mass, therefore 1g starch in 100g water. Water has a density of 1g/ml, therefore 1g starch in 100mL water. Measure 100ml, 200 ml, 300ml and 400 ml into a series of standard flask labeled 1, 2, 3 and 4. Add water to all 4 standard flask (1,2,3 and 4) to bring the total volume to 1000 ml.

8.2.4 Preparation of 0.1 M Iodine solution

Measure out 3.0 g of potassium iodide (KI) into an appropriate beaker. Moisten the potassium iodide with a few drops of water. Measure out 2.54 g of iodine and add to the moistened potassium iodide. Add a small volume of water and stir. Pour the solution into a graduated cylinder and dilute with distilled water to the final volume of 100mL. If there are any bits of iodine remaining, return the solution to the beaker and leave it on a magnetic stirrer for several minutes.

8.2.5 Malted Wheat flour

A commercial sample of wheat was germinated, dried at 100°C for 12 hours and was

- powdered using a mixer into whole wheat flour.

8.2.6 Extraction of crude alpha amylase

A 0.2% Calcium acetate solution was used as the extraction medium. Germinated whole wheat flour (50g) was added slowly with stirring to 1000ml beaker. Stirring was continuedfor 2 hours after addition of flour was completed. The suspension was centrifuged at 4000rpm for 20 minute. The resulting clear solution was stored at 2 to 3°C. Since beta amylase interferes with the enzymatic determination of alpha amylase used in the study, it was inactivated by heating the extract at 70°C for 15 minute. Alpha amylase between the pH 6.5 to 8 is resistant to inactivation by this treatment. In addition to beta amylase many other heat labile protein are denatured by this operation. The extract was first adjusted pH to 6.6 with ammonium hydroxide. Heat treatment was carried out in two 1 liter beaker, each heated by hot plates equipped with magnetic stirrers.

8.2.7 Yeast Extract Peptone Dextrose Medium (YEPD or YPD broth)

YPD Broth (YEPD Broth) is recommended for the growth of Saccharomyces cerevisiae for molecular biology purposes.20grams of bacteriological peptone,10grams of Yeast extract and 20grams of dextrose were taken and suspended in 1000 ml distilled water. The medium was dissolved completely by using magnetic stirrer and heater at temperature 72˚C. The sample was sterilized by autoclave at 15 lbs pressure (121°C) for 15 minutes. Cultural characteristicsobserved after incubation at 25-30°C for 48-72 hours and pH was maintained at 6.5 to 6.7 ranges.

Fig. 8.12.1 Medium before heating Figure 8.12.2 Medium after heating using temperature controlled magnetic stirrer

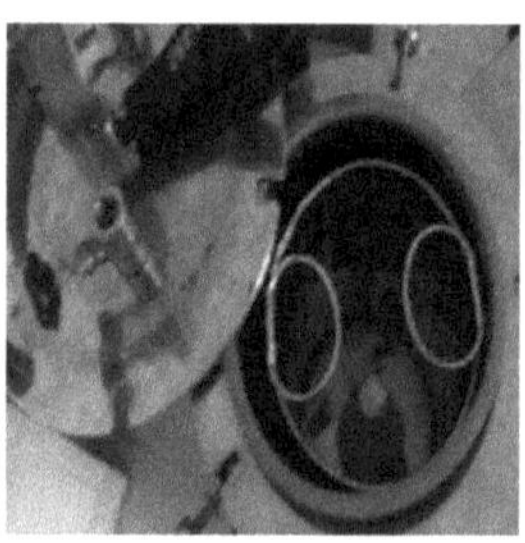

Fig. 8.13.1 Autoclave Fig. 8.13.2 Autoclaved medium

Fig. 8.14 Yeast cultivated in YEPD medium

8.2.8 Potassium dichromate solution

34 g of potassium dichromate was dissolved in 500 ml distilled water in 1L flask. 325 ml concentrated sulfuric acid was added slowly keeping the flask in ice

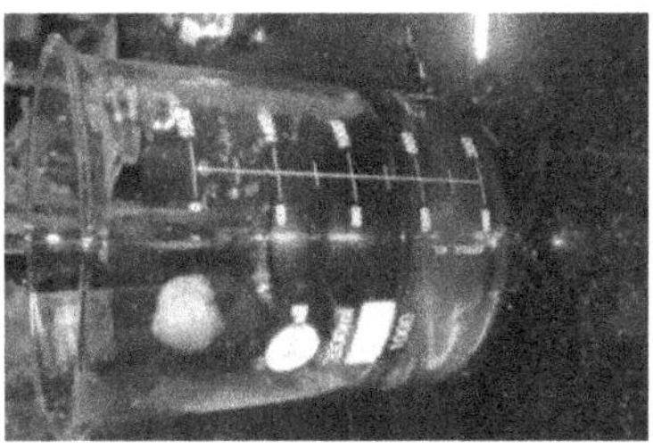

Fig. 8.15 Potassium dichromate solution

Rests of the chemicals used were 0.2N H2SO4, Concentrated Sulfuric acid, 100% ethanol, distilled water, 0.1N Hydrochloric acid, glucose, 0.1N Sodium hydroxide, Sodium thiosulfateanhydrous, Silica gel and Activated carbon.

8.2.9 Moisture Content

A 1000 ml plastic beaker was taken and weighed.500g of duckweed was weighed and transferred to a metal weighing pan .The duckweed was dried at 100˚C for 24 hours in hot airoven.The sample was reweighed and then substracted the weight of container. Dried weight of the sample is 41 g. The moisture content was determined by the equation

$$\mathbf{M = ((W_w - W_d)/W_w)*100}$$

In which

M = Moisture content (%) of duckweed

W_w =Wet weight of the sample and

W_d = Weight of the sample after drying

8.3 Determination Of Starch Concentration By Spectrophotometry

8.3.1 Extraction of total available starch from duckweed

The fresh Lemnaceae was dried in a hot air oven at 100˚C for 24 hour. The dried duckweed was powdered and dried at 70° to constant weight. The Soxhlet extractor is a master of the slow liquid-solid extraction, a procedure used to separate the poorly soluble from the insoluble. During use, the inner core of the soxhlet is stuffed with dried duckweed that has been wrapped in a muslin cloth. The extracting solvent that is 0.2N H_2S0_4 is heated to reflux in a round bottom flask below and the condensate drips down into the filter. As the hot solvent fills the extractor the compound of interest slowly dissolves, forming a low molarity solution. When the reactor is full a siphon is triggered, and the solvent flows down to the Round bottom flask. This cycle repeats for up to several cycles, removing all trace of the soluble compound from the upper chamber. The hot solution was filtered through Whatman ilter paper. Proteins were not removed. The filtrate was cooled, neutralized with NaOH solution and diluted to volume with distilled water.1 drop of iodine solution was added and mixed thoroughly. UV spectrophotometer was used to record the absorbance of solution at 610 nm.

10 ml of standard solutions were taken and 1 drop of iodine solution was added .The standardsolutions were analyzed by using UV spectrophotometer at 610nm and calibration curve was plotted. From the calibration curve concentration of starch in sample was determined.

8.4 LIQUEFACTION

Liquefaction is the hydrolysis of starch to oligosaccharides. This is done by holding starch residue at 105°c for 7 minutes at pH 6-6.5 in pressure stable alpha amylase enzyme. Small quantity of calcium salt is also added to help stabilize the enzyme. During the 7 minutes starch hydrates and is broken by shearing force in the autoclave and by the action of enzyme. Starch + water Oligosaccharides (3.2)

After the initial liquefaction starch is cooled to 97°C and transferred to multi chamber reactor where the solution is held for 90 minutes to reach dextrose equivalent of 10-12 units. As the name implies liquefaction lowers the viscosity of the solution.

8.5 SIMULTANEOUS SACCHARIFICATION AND FERMENTATION (SSF) PROCESS

8.5.1 Saccharification

Liquefied sample was taken in a 1000ml beaker .The yeast cultivated over the incubated yeast extract peptone broth was added to the sample. Temperature was maintained at 30 °C and agitation was maintained at 400 rpm. pH was controlled at 5.5. Inoculum was grown for 22 h at 30 °C. Samples were taken every two hour during first 8 hours of saccharification andthen after 12 hour.

8.5.2 Fermentation

After inoculation, 120 hours fermentation was carried out .The samples were taken in 250ml conical flask with stopper and placed in shaker inside a BOD incubator cum shaker at 30°C for 120 hours. The samples were taken every 24 hours of fermentation time and analysis was done.

8.6 ANALYTIC METHOD

8.6.1 DETERMINATION OF BIOMASS

For biomass determination, 5 ml of fermentation mash was taken out and centrifuged at 4000rpm for 6 minutes. The cells were suspended in 5 ml distilled water .The absorbance was determined at 600nm with Ultraviolet Visible spectrophotometer. If the absorbance is above 1, the sample was diluted until absorbance is less than 1.

8.6.2 Dry Weight Measurement:

100 ml of fermentation mash was taken out and dried in an empty weighing pan inside a hot air oven. The broth was centrifuged at 4000 rpm for 10 minutes. The clear broth wasdiscarded and the cell paste was scrapped from the centrifuge tube into a weighing pan. The centrifuge tube was rinsed with a few ml of water. The rinsed water was poured into the weighing pan .The cell paste was dried in a hot air oven set at 100°C. The weight of the pan plus the cell paste was weighed periodically until there is no further decrease in the dry weight. It had taken 24 hours to dry the sample completely, depending on the oven temperature and the thickness of the paste. Calculated the difference in the weight and thedry weight is expressed in g/l. The sample was diluted by using 250 ml distilled water and the absorbance of the sample was measured with a spectrophotometer at 550 nm. Other wavelengths may also be used, but one must be consistent. Generate a calibration curve to relate the absorbance with cell dry weight.

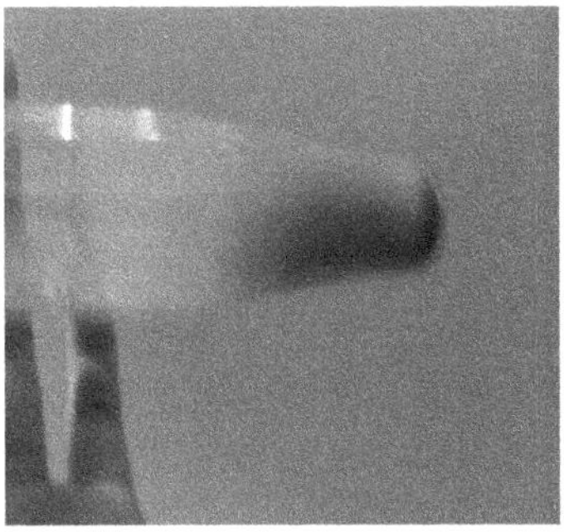
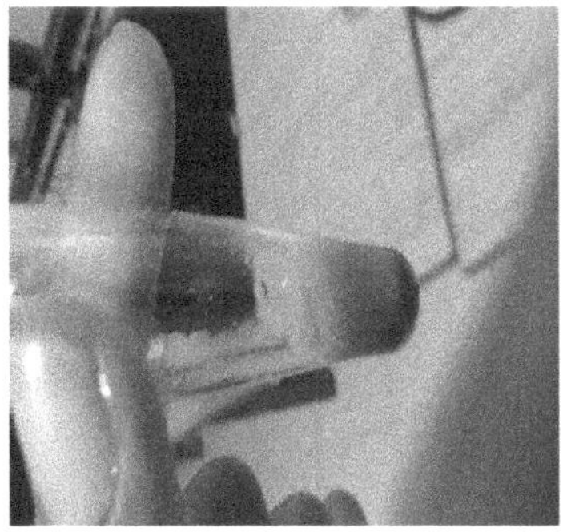

Fig. 8.16 Centrifuged biomass sample figure 3.15 Cell paste

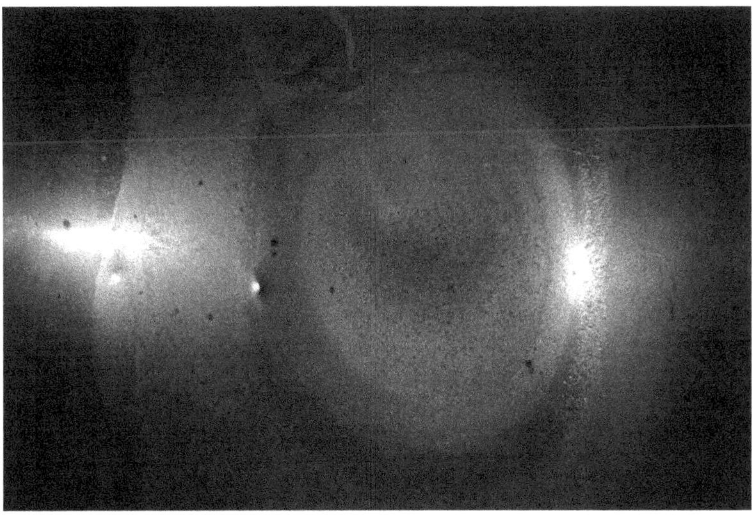

Fig. 8.17 Dried Cell paste

8.7 DETERMINATION OF REDUCING SUGAR

For the determination of the reducing sugar (S) concentration, the fermentation broth was taken and centrifuged, and then 20 μL was diluted to 25mL with distilled water and mixed with 3 ml concentrated sulfuric acid and determined with 520 nm in UV spectrophotometry. Eight standard solutions were prepared with concentration of glucose varying from 1000 to 7000 mg/l .Same procedure for sample was repeated and using UV spectrophotometer absorbance was measured at 520 nm.

8.8 DETERMINATION OF ETHANOL CONCENTRATION

Principle

In the presence of sulfuric acid, alcohol forms a complex with potassium dichromate which has an absorption maximum at 600 nm.

8.8.1 Preparation of standard curve

1,5,10,30,50,90 % v/v alcohol was prepared in various test tubes.10 ml of various concentration of alcohol was taken in 250 ml distillation flask. 15 ml of distillate was distilled and collected in 50 ml volumetric flask containing 25 ml of potassium dichromate. Then solution was made up to 50 ml. The alcohol potassium dichromate complex was kept at60^0 C for 30 minutes. Optical density was measured at 600 nm. A standard graph was plotted with concentration of alcohol on X axis and optical density at 600 nm at Y axis.

8.8.2 Determination of Ethanol concentration of sample

10 ml of sample was mixed with 250ml distilled water .Diluted sample was taken in a 500 mlround bottom flask and distilled .The distilled sample was transferred into 250ml conical flask containing 25 ml of potassium dichromate. Then solution was made up to 50 ml. The ethanol potassium dichromate complex was kept at 60^0 C for 30 minutes. Optical density wasmeasured at 600 nm. The alcohol concentration was found from graph.

Fig. 8.18 Ethanol potassium dichromate complex on water bath

8.9 AZEOTROPIC DISTILLATION

100ml of sample was transferred to 250ml round bottom flask. The glass beads were added to round bottom flask so that contamination of distillate was avoided and glassware was protected from sudden pressure wave. A drop of glycerol was placed on the bulb end of the thermometer and carefully inserted the thermometer into one hole of the stopper. Thermometer was pressed gently with twisted motion. Bulb end of the thermometer protruded a few cm below the bottom of the stopper. The 250 ml round bottom flask was placed on the hot water bath at 95°C.The bend was connected to the top of the round bottom flask. A condenser was connected to the free end of the bend. The 100 ml graduated cylinder was placed on the free end of the condenser. As the liquid got boiled the temperature of the thermometer was increased to indicate the temperature of the vapor that was filled in the flask. When temperature of still head was 56°C, the vapor temperature of boiling acetone, liquid started boiling. As the temperature reached 60°C in thermometer distillate was collected. The temperature was stayed at 60°C until acetone has all boiled away and there is none of it left in the solution. The distillate collection was stopped when the temperature got increased from 60°C.The distillation was stopped and bottom product was transferred to a closed tight container and placed inside refrigerator.

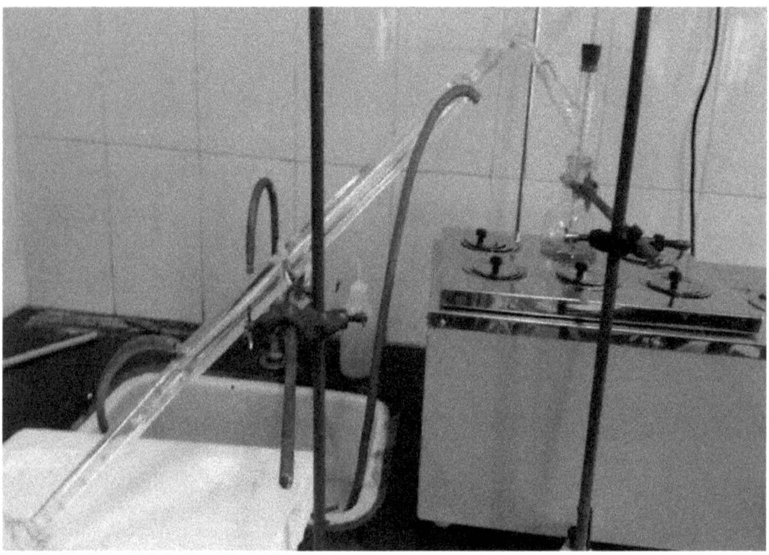

Fig. 8.19 Azeotropic distillation experimental setup

8.10 SOLVENT EXTRACTION FROM DISTILLATE

The solvent collection flask of the unit should always be emptied prior use to preventaccidentally mixing of incompatible chemicals. The flask with the solution was placed on the rotary evaporator. A metal or Keck clip was used to secure the flask. The dial on the motor was used for speed control of the flask rotation. A typical rotavap was used with a variable speed sparkless induction motor that spins at 0-220 rpm and provided high constant torque. Here it was set at 8 rpm. The aspirator vacuum was turned on.The flask was lowered into the water bath. A convenient handle (with height locking mechanism) was used to move the entire condenser/motor/flask assembly up and down. Often the tilt of the condenser assembly can also be adjusted. The water bath temperature should not exceed the boilingpoint of the solvent .The solvent acetone was collected on the condenser at 50.8°C and dripped into the receiving flask. Once all the solvent evaporated, the vacuum was released,.The flask was raised out of the water bath and the spinning was discontinued. The receiving flask was emptied upon completion of the evaporation.

8.11 DEHYDRATION

19ml of Ethanol sample obtained was taken. Three 250 ml conical flask was taken and 5 mlof sample was transferred to each flask. To the conical flask A,B and C 0.3 g of silica gel, sodium thiosulfate anhydrous and activated carbon was added .The samples was shaken for 10 minutes in a shaker and the shaker samples was filtered using whatman filter paper. The concentration of ethanol in shaked samples was determined by using UV spectrophotometer.

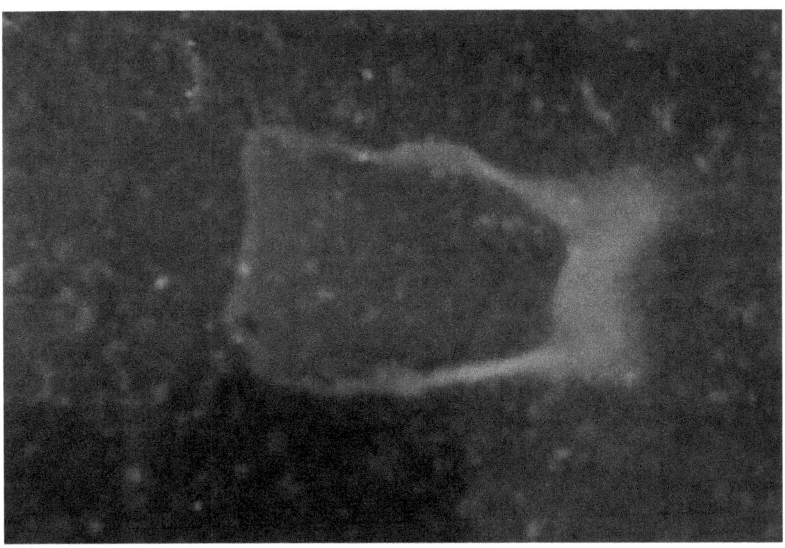

Fig. 8.20 Ethanol burning with blue flame

8.12 PARAMETER ESTIMATION FOR FERMENTATION USING MMATLAB

The rate equations for this process are given below:

$$\frac{dS}{dt} = r_S = -\left(\frac{\mu_{max} S}{Y_{XS}(K_S + S)} + m_S \right) X$$

$$\frac{dP}{dt} = r_P = -\left(Y \frac{\mu_{max} S}{K_S + S} + m \right) X$$

$$\frac{dx}{dt} = \frac{\mu_{max} S}{K_S + S} X$$

8.12.1 Computation

Substrates + Cells ➔ Extracellular Products + More Cells

ΣS + X ➔ ΣP + nX

Exponential growth rate is first order with respect to X

$\frac{dX}{dt} = \mu X$ $\qquad$ $X = X_0$ at $t = 0$

Where X and X_0 are cell concentration at t and t=0.

Integration yields $\ln(X/X_0) = \mu(t - t_0)$

8.12.2 Doubling time (time required to double microbial mass or number). t_d

$$t_d = \frac{\ln 2}{\mu}$$

8.12.3 Specific growth rate:

Ignored the lag period of the cultivation where in no increase in biomass was observed. $\log_e$ "X" versus "t" curve was plotted in the exponential phase of the growth. The slope of this curve was the specific growth rate of the culture.

8.12.4 Maximum specific growth rate and saturation constant:

"X" and "S" versus time curves were plotted. 5 points in the declining phase on the growth curve was selected (reducing substrate concentration). Tangents were drawn for "X" versus "t" curve and calculated "dX/dt". The corresponding values of "S" were identified at the samepoints which were used for making the tangents. "1 / μ" versus "1 / S" graph was plotted. A linear regression line through the data points was fitted.

Assumed the applicability of Monod's Equation

$\mu = \mu_{max} S / (K_S + S)$

$1 / \mu = (K_S / \mu_{max}) 1 / S + 1 / \mu_{max}$

The slope of line = (K_S / μ_{max}) and the intercept $(1/ \mu_{max})$ on abscissa = $(1/\mu_m)$.From these compute the values of "μ_m" and "K_s".

8.12.5 Overall productivity:

To compute overall productivity of biomass calculate $(X_F - X_O) / (T_F - T_O)$.

8.12.6 Maintenance Coefficient

Maintenance Coefficient = specific rate of substrate uptake for cell maintenance:

8.12.7 Yield

Observed or apparent yield is the ratio of mass or moles of product present to total mass ormoles of reactant consumed.

8.12.8 Yield coefficient

Growth Yield in fermentation is: $Y_{X/S} = \frac{-DX}{DS}$

Apparent Growth Yield at end of a fermentation (batch growth period)

$$Y_{P/S} = \frac{-DP}{DS}$$

$Y_{P/S}$ = mass or moles of product formed per unit mass or moles of substrate consumed

$Y_{P/X}$ = mass or moles of product formed per unit mass or moles of biomass

8.12.9 MATLAB

MATLAB(matrix laboratory) is a multiparadigm numerical computing environment and fourth-generation programming language. Apropprietary programming

language developed by MathWorks, MATLAB allows matrix manipulations, plotting of functions and data, implementation of algorithms, creation of user interfaces, and interfacing with programs written in other languages including C, C++,Fortran and Python.The version (8.1.0.604) of MATLAB R2013a Student Version (32-bit) was used here.

8.13 RESULTS AND DISCUSSION

8.13.1 MOISTURE CONTENT

In which W_w=500g

W_d= 41g

M = Moisture content (%) of duckweed=91.8%

8.13.2 TO FIND PERCENTAGE OF STARCH CONTENT IN DUCKWEED

Starch extracted using Soxhlet extractor = 19.48g

Dry weight of duckweed = 41g

Starch content in duckweed = 47.51%

8.13.3 DETERMINATION OF STARCH CONCENTRATION

The starch sample was analysed by UV apectrophotometer .The obtained absorbance of the starch sample is 1.63.The concentration of starch in solution is 7 g/l.

8.13.4 FERMENTATION

During the duckweed fermentation, reducing sugar, ethanol production, and its biomass formation was plotted against fermentation time. Consequently, it is necessary to study the kinetic of cell growth at 30°C temperature.

8.13.5 Cell Growth at Various Temperatures

Saccharomyces cerevisiae cell growth showed a classical cell growth trend under 30°C temperature. The batch fermentation process of duckweed can be separated into different stages. The first stage is the lag phase. In this stage, the biomass increased slowly, and then the biomass increased faster in the subsequent exponential growth phase. Finally, 6 g/l of biomass was attained at 30°C in the stationary phase. Reducing sugar formation began to form as soon as the fermentation started. However, ethanol production was initiated when theyeast cells reached the exponential phase and cell growth occurred simultaneously. Similar phenomenon has been reported.

Biomass Concentrationin g/l

Biomass Concentration Versus time

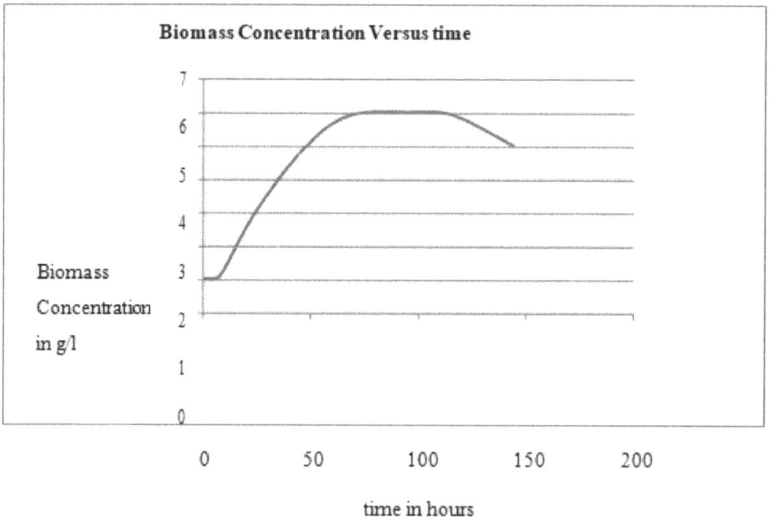

Fig. 8.21 Variation of biomass concentration with time

8.13.6 Determination of reducing sugar

The glucose concentration of the sample during the time of saccharification was measured using UV spectrophotometer. The glucose concentration determined was plotted against time. A maximum reducing sugar concentration of 6.2 g/l was accumulated in the fermentation broth at 22 h under 30°C. The profile of the reducing sugar indicates that during the initial stage the saccharification rate was higher than the fermentation rate leading to the accumulation of the reducing sugar. The yeast was in the lag phase with lower consumption of reducing sugar for cell growth. Subsequently, the yeast entered the exponential stage during which reducing sugar was consumed.

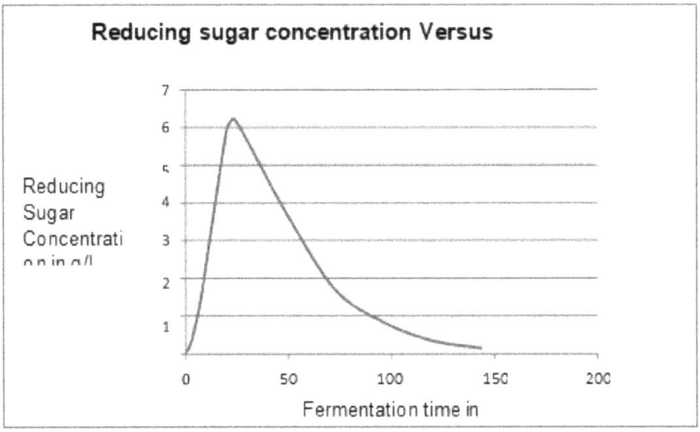

Fig. 8.22 Variation of reducing sugar concentration with time

8.13.7 Determination of ethanol concentration

Ethanol was the metabolic product of Saccharomyces cerevisiae fermentation. Ethanol is assumed to be directly proportional to the dry cell weight and substrate as well as reducing sugar concentration. Concentration of ethanol in the sample solution collected during every 24 hours time interval was determined from standard ethanol curve. The ethanol concentration of sample was plotted against fermentation time.

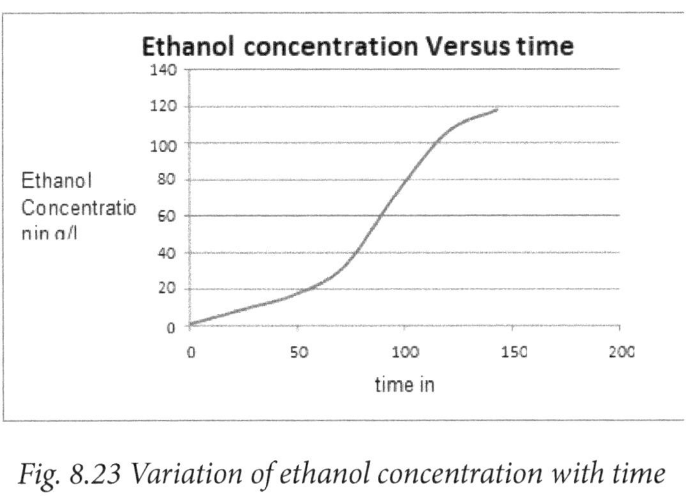

Fig. 8.23 Variation of ethanol concentration with time

8.14 AZEOTROPIC DISTILLATION

Ethanol sample obtained after azeotropic distillation was analyzed by UV spectrophotometer at 600nm.The absorbance was determined and the value of absorbance was 2.069.Corresponding concentration was found out from standard ethanol graph .95% ethanol was obtained after azeotropic distillation using acetone.

8.15 DEHYDRATION

Dehydration was carried out by various drying agent such as silica, sodium thiosulfate anhydrous and activated carbon. The 3 ethanol samples obtained after drying was analysed using UV spectrophotometer. The absorbance obtained for silica, sodium thiosulfate anhydrous and activated carbon was obtained as 2.110, 2.131 and2.116.corresponding ethanol concentration was determined from standard graph and percentage of ethanol content in silica dried ethanol was 97%,sodium thiosulfate dried ethanol was 99% andactivated carbon dried ethanol was 98%.Among this three drying agent sodium thiosulfate anhydrous was found as best for drying ethanol water mixture.

8.16 ESTIMATION OF GROWTH KINETIC PARAMETERS IN BATCH FERMENTATION

The specific growth rate, glucose consumption rate, production rate and the yield coefficient are commonly adopted to assess microbial performance.

8.16.1 Specific growth rate:

$$\frac{dX}{dt} = \mu X$$

The proportionality constant μ is called the specific growth rate.μ is positive for growth phase.μ is a function of nutrient supply is changing with time.

Logarithm of biomass concentration versus fermentation time graph is a straight line with slope 0.011. The slope of this curve is the specific growth rate of the culture. So specific growth rate is 0.011(μ = 0.011 h^{-1}).

8.16.2 Maximum specific growth rate and saturation constant:

Saturation constant is 50.59 $\frac{l}{g}$ and Maximum growth rate obtained is 0.3666 h^{-1}.

The idea of microbial growth kinetics has been dominated by an empirical model originally proposed by Monod. The Monod model introduced the concept of a growth limitingsubstrate.

$$\mu = \mu_{max} \frac{S}{K_S + S} \qquad (4.3)$$

Where μ = specific growth rate, μ_{max} = maximum specific growth rate, S = substrate concentration, K_S = substrate saturation constant (i.e. substrate concentration at half μ_{max}). In Monod's model, the growth rate is related to the concentration of a single growth-limiting substrate through the parameters μ_{max} and K_S.The slope of the Monod's model plotis (K_S / μ_{max}) and the intercept on abscissa is ($1/\mu_{max}$).

8.16.2 Overall productivity

The ratio of the total biomass produced out of the entire substrate to entire time period.

Overall productivity obtained is 0.064 .

8.16.3 Doubling time ,t_d

Doubling time calculated is 63.013h. Doubling time is the time at which $X=2X_0$.

8.16.4 Maintenance Coefficient

Substrate maintenance rate constant calculated is 0.0117 g substrate/g biomass/h and Productmaintenance rate constant is 0.018 g product/g biomass/h.

The rate of substrate consumed in order to maintain cells is proportional to the mass density. The higher the value of the maintenance coefficient, the more substrate goes towards maintaining the culture without contributing to cell mass or product formation.

m is the maintenance coefficient expressed as a substrate demand to maintain cell viabilityper unit of cells per time

8.16.5 Yield Coefficient:

Growth Yield $Y_{X/S}$ in fermentation is 0.711 g biomass/g substrate and Yield on product, $Y_{P/X}$ is 7.75 g product/g biomass

Cell mass and product formation by microorganisms can be described quantitatively by yield coefficients expressed as the mass of cells or product formed per unit mass of substrate consumed, $Y_{X/S}$ and $Y_{P/S}$ for cells and product, respectively.

8.16.6 Yield

$Y_{P/S}$,yield calculated is 5.50 g of ethanol /g of glucose.

Yield is the ratio of mass or moles of product present to total mass or moles of reactantconsumed.

8.16.7 MATLAB Plotting

The MATLAB program was also employed to solve the three ordinary differential equationsderived from material balance of cell, glucose and ethanol, with the data obtained from the experiments

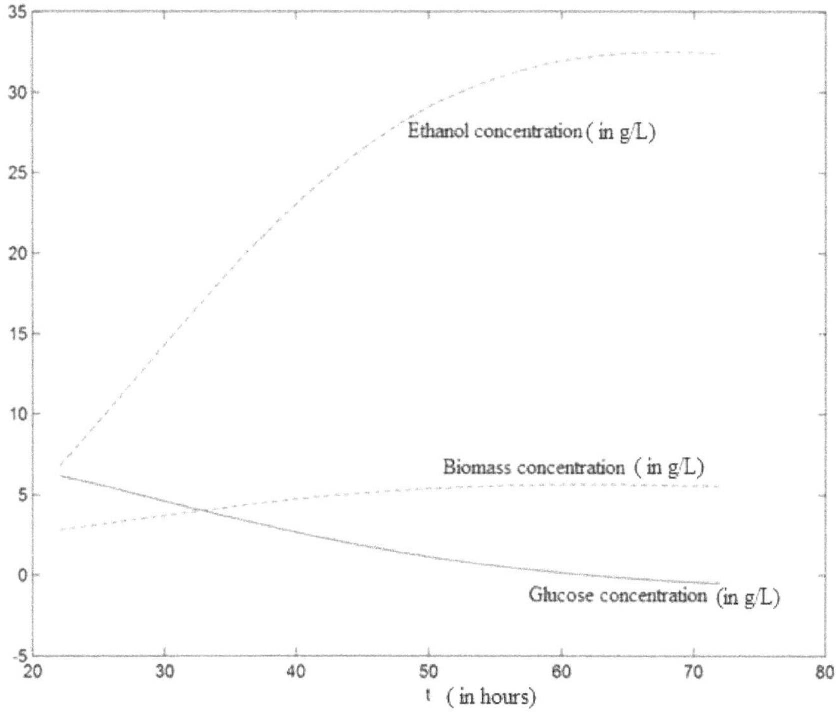

Fig. 8.24 Graph obtained in MATLAB after executing the program

CONCLUSIONS

Enzymatic saccharification was performed for alpha amylase pretreated duckweed starch. The starch yield in duckweed was obtained as 47.51%.After liquefaction and saccharification, glucose concentration was increased to 6.2 g/l. A kinetic model was employed for studying the fermentation of sugars produced by saccharification of duckweed using

Saccharomyces cerevisiae. Based on the Monod equation the growth of yeast, growth associated production of ethanol and utilization of sugars for biomass production and maintenance were estimated using MATLAB R2013a software. The kinetic parameters such as $Y_{P/S}$, $Y_{P/X}$, $Y_{X/S}$ and m were found out.

- Growth Yield in a fermentation $Y_{X/S}$ = 0.711 g biomass/ g substrate
- Yield on product $Y_{P/X}$ = 7.75 g product/ g biomass
- Substrate maintenance rate constant, m_s= 0.0117 g substrate/g biomass/h
- Product maintenance rate constant m_P = 0.018 g product/g biomass/h
- Yield = $Y_{P/S}$ = 5.50 g of product /g of glucose

The ethanol concentration of fermented sample was found to be 15%. It was concentrated to 95% by using azeotropic distillation. Sodium thiosulfate anhydrous , silica gel and activated carbon was used to dry the sample obtained after distillation and sodium thiosulfate anhydrous was found more effective than other two drying agent.

FUTURE PERSPECTIVES

To improve the application of duckweeds for biofuel production, further studies are needed into duckweed selection, optimization of biomass production, enhancement of starch accumulation and use for production of various biofuels. When duckweed is used forbioethanol production, it is critical that the duckweed has both high biomass production and high starch content. To increase starch content, enrichment of CO_2 in growth media couldstimulate photosynthesis and carbohydrate production. Accumulation of starch could also be achieved through the inhibition of starch consumption in cell metabolism. An understandingof starch degradation in cells is very important for further studies of starch accumulation in duckweed. Duckweed fronds are similar to higher plant leaves, so information on starchdegradation in such leaves could be important for exploring transitory starch degradation in duckweed fronds. A fundamental understanding of transitory starch degradation will be essential in developing strategies to inhibit starch degradation and thus improve starchaccumulation in duckweed.

CHAPTER 9

INDUSTRIAL VISIT TO TEXTILE INDUSTRY AND SUSTAINABLE SOLUTIONS

9.1 Introduction

Industrial visits provide valuable practical exposure to real-time manufacturing operations, industrial management practices, environmental protection systems, and quality assurance techniques. They bridge the gap between theoretical learning and industrial implementation by enabling students and researchers to observe actual production environments and understand large-scale industrial processes.

The industrial visit to a textile processing industry offered comprehensive insight into modern textile manufacturing operations, including:

- Yarn processing
- Fabric dyeing
- Chemical treatment
- Quality testing
- Resource management
- Wastewater treatment
- Sustainable manufacturing practices

The visit also highlighted major environmental challenges associated with textile industries such as:

- High water consumption
- Chemical-intensive processing

- Wastewater generation
- Energy consumption
- Synthetic material dependency

Based on industrial observations, sustainable solutions involving process optimization, agro-waste utilization, circular economy principles, and Life Cycle Assessment (LCA) approaches were identified and proposed.

9.2 Objectives of the Industrial Visit

The industrial visit was conducted with the following objectives:

- To understand textile manufacturing operations
- To study yarn and fabric dyeing processes
- To observe industrial chemical treatment methods
- To analyze wastewater treatment practices
- To evaluate quality assurance systems
- To identify environmental issues in textile industries
- To explore sustainable alternatives and circular economy strategies

9.3 Overview of Textile Manufacturing Operations

The textile manufacturing process involved several interconnected stages starting from raw material preparation to final packaging and export.

9.3.1 Raw Material Handling and Initial Processing

The manufacturing process began with the inspection and preparation of cotton yarn and fabric materials.

Yarn Inspection

The yarn underwent detailed quality inspection to ensure:

- Uniformity
- Strength
- Defect-free processing
- Proper dye absorption

Chemical Processing Sections

The facility consisted of:

- 13 specialized chemical processing and dyeing sections

These sections enabled efficient handling of:

- Multiple fabric types
- Different color shades
- Large-scale production operations

Production Capacity

The industry demonstrated large-scale processing capability:

- Approximately 5 tonnes of yarn dyed daily
- Nearly 5000 meters of fabric processed daily

Color Production System

The plant utilized:

- Three binary color systems

to produce:

- Approximately 2000–5000 color shades

This enabled flexibility in meeting diverse market demands.

Manufacturing Flow

The process sequence included:

- Yarn dyeing
- Fabric weaving
- Quality grading
- Finishing operations
- Packaging and dispatch

9.4 Chemical Processing and Dyeing Techniques

Chemical processing is one of the most important stages in textile manufacturing because it determines:

- Color quality
- Fabric appearance
- Texture
- Durability

9.4.1 Purification and Bleaching

Cotton fabrics were first purified to remove:

- Natural impurities
- Oils
- Waxes
- Dust particles

Bleaching created a uniform white base suitable for dyeing.

9.4.2 Softening Treatment

Softening agents were applied after bleaching to improve:

- Fabric softness
- Surface smoothness
- User comfort

9.4.3 Reactive Dyeing Process

Reactive dyes were primarily used due to:

- Strong fiber bonding
- Excellent color fastness
- Better washing resistance

9.4.4 Chemical Treatments

Several chemical treatments were involved:

- Alkali treatment
- Acetic acid treatment
- Neutralization process

These treatments improved:

- Dye fixation
- Fabric stability
- Color consistency

9.4.5 Auxiliary Chemicals

Chemicals such as:

- Soda ash
- Salts
- Alkaline substances

were added to promote:

- Uniform dye penetration
- Permanent color fixation

9.4.6 Soaping Process

The soaping step removed unfixed dyes to:

- Prevent color bleeding
- Improve wash durability
- Enhance fabric appearance

9.4.7 Final Softening

Final softening improved:

- Hand feel
- Texture
- Surface finish

9.5 Yarn and Fabric Dyeing Operations

9.5.1 Yarn Dyeing

Grey cotton yarn was loaded into specialized cages for wet processing operations.

Processing Stages

- Water treatment
- Dyeing
- Water extraction
- Drying

Radio Frequency Dryers

Advanced radio frequency dryers were used for:

- Rapid moisture removal
- Uniform drying
- Energy-efficient processing

9.5.2 Fabric Dyeing

Fabric dyeing involved:

- Singeing operations
- Hydro extraction
- Controlled drying

Singeing Machine

The singeing machine removed:

- Loose surface fibers
- Fuzz particles

This improved:

- Fabric smoothness
- Dye uniformity

9.5.3 Post-Dyeing Handling

After dyeing:

- Fabrics were folded carefully
- Pinned to prevent creases
- Prepared for packaging

9.6 Environmental Management – Effluent Treatment Plant (ETP)

Textile industries generate large volumes of wastewater containing:

- Dyes
- Salts
- Suspended solids
- Organic pollutants

To reduce environmental impact, the industry employed advanced wastewater treatment systems.

9.6.1 Wastewater Volume

The facility treated:

- Approximately 25 million liters of wastewater daily

9.6.2 Wastewater Treatment Process

The treatment stages included:

- Preliminary treatment
- Anaerobic digestion
- Multi-stage filtration
- Activated carbon treatment
- Water recycling

9.6.3 Filtration Systems

Advanced filtration methods included:

- Membrane filtration
- Activated carbon filtration

These systems removed:

- Dye particles
- Dissolved solids
- Organic contaminants

9.6.4 Water Quality Monitoring

Important parameters monitored included:

- Chemical Oxygen Demand (COD)
- Total Suspended Solids (TSS)

to ensure compliance with environmental regulations.

9.6.5 Water Reuse

Treated water was recycled for:

- Washing operations
- Industrial reuse

This reduced:

- Freshwater consumption
- Environmental discharge

9.7 Resource Management and Process Efficiency

9.7.1 Water Consumption

The textile dyeing process consumed:

- Approximately 80 kg of water per kg of dye applied

This highlighted the need for improved water management.

9.7.2 Energy Management

The industry maintained:

- Controlled electricity usage
- Efficient processing systems

to reduce operational costs.

9.7.3 Dye Bath Management

Multiple dye baths consumed significant quantities of:

- Water
- Chemicals
- Energy

requiring careful optimization

9.7.4 Preparatory Operations

Cheese winding ensured:

- Proper yarn arrangement
- Uniform dye absorption

9.8 Quality Control and Laboratory Testing

The textile industry maintained strict quality control procedures to ensure product performance and customer satisfaction.

9.8.1 Sample Testing Conditions

Testing was conducted under controlled:

- Temperature
- Humidity

to ensure consistent results.

9.8.2 Washing Performance Tests

Fabrics were evaluated under various washing conditions to assess:

- Shrinkage
- Color stability
- Durability

Fig. 9.1 Dying and washing unit

9.8.3 Microscopic Analysis

Microscopic examinations detected:

- Fiber defects
- Structural irregularities
- Surface damage

9.8.4 Color Fastness Testing

Color durability was tested under:

- Heat exposure
- Washing conditions
- Frictional stress

9.8.5 Fabric Durability Tests

Tests included:

- Abrasion resistance
- Pilling resistance
- Tensile strength measurement

9.9 Final Processing and Packaging

Final operations ensured:

- Product quality
- Export readiness
- Safe transportation

9.9.1 Inspection

Final inspection checked:

- Color consistency
- Fabric defects
- Dimensional accuracy

9.9.2 Packaging

Fabrics were:

- Folded carefully
- Pinned properly
- Packed according to export standards

9.10 Problem Statement

The industrial visit revealed several environmental and operational challenges within textile processing industries.

- Major concerns included:
- High water consumption
- Extensive chemical usage
- Wastewater generation
- Synthetic material dependency
- Energy-intensive operations

The observed facility processed:

- 5 tonnes of yarn daily
- 5000 meters of fabric daily
- Generated nearly 25 million liters of wastewater

Despite the presence of Effluent Treatment Plants (ETPs), maintaining acceptable:

- COD levels
- TSS values

remained challenging.

Simultaneously, synthetic petroleum-based textile and leather materials contribute significantly to environmental pollution.

Large quantities of agro-industrial waste such as:

- Rice processing wastewater
- Nutmeg rind waste
- Agricultural biomass

remain underutilized despite possessing valuable industrial potential.

Therefore, there is a strong need for:

- Sustainable textile processing
- Agro-waste utilization

- Resource-efficient manufacturing
- Circular economy implementation
- Life Cycle Assessment (LCA)-based environmental management

9.11 Life Cycle Assessment (LCA) Approach

To evaluate environmental impacts systematically, a simplified Life Cycle Assessment framework was proposed.

The LCA framework considered:

- Raw material input
- Water usage
- Energy consumption
- Waste generation
- Wastewater treatment
- Product output

9.11.1 Key Environmental Hotspots

The study identified major hotspots including:

Dyeing operations

Washing processes

Chemical treatment stages

These operations contributed significantly to:

- Water consumption
- Chemical discharge
- Energy usage

9.11.2 Role of ETP

The Effluent Treatment Plant played a critical role in:

- Pollution reduction
- Water recycling
- Environmental protection

Fig. 9.2 Waste Water treatment using biological organism

Fig. 9.3 Filter press for water treatment

9.12 Proposed Sustainable Solutions

9.12.1 Process Optimization

- Optimization of pH and temperature
- Reduced salt and alkali usage
- Improved dye fixation efficiency
- Optimized washing operations

9.12.2 Water Conservation and Recycling

- Use of partially treated effluent for initial washing
- Increased recycled water utilization
- Reduced water intensity during dyeing

9.12.3 Improvement of ETP Systems

Enhancement of:

- Biological treatment systems
- Carbon filtration
- Membrane separation
- COD and TSS monitoring

9.12.4 Utilization of Sustainable Biomaterials

Replacement of petroleum-based materials with:

- PLA biopolymers
- Bio-pectin composites
- Agro-waste-based vegan leather

9.12.5 Integration of LCA and LCI

- Development of:

- Life Cycle Inventory databases
- Environmental monitoring systems
- Sustainability assessment tools

9.12.6 Circular Economy Strategies

- Process stream reuse
- Agro-waste utilization
- Biodegradable product development
- Sustainable waste management

9.13 Integrated Sustainable Approach

The project proposed an integrated sustainability strategy combining:

- Sustainable material development
- Textile process optimization
- Circular economy implementation
- Life Cycle Assessment

9.13.1 Bio-Based Vegan Leather Development

The proposed material involved:

- PLA derived from rice wastewater
- Bio-pectin extracted from nutmeg rind waste

This reduced:

- Petroleum dependency
- Synthetic material usage

9.13.2 Process-Level Improvements

Optimization of:

- Chemical dosage
- Temperature
- Water management
- Energy utilization

could significantly improve sustainability performance.

9.13.3 Development of Life Cycle Inventory

The project emphasized collecting data related to:

- Material input
- Water usage
- Energy consumption
- Emissions
- Waste generation

9.13.4 Circular Manufacturing System

The integrated system promoted:

- Waste reuse
- Biodegradable materials
- Compostable products
- Sustainable end-of-life management

CONCLUSION

The industrial visit provided detailed practical exposure to textile manufacturing operations, environmental management systems, quality control procedures, and resource utilization practices. The observations highlighted both the technological advancements and environmental challenges associated with modern textile industries.

The study identified significant opportunities for improving sustainability through:

- Process optimization
- Water conservation
- Agro-waste utilization
- Bio-based material development
- Circular economy implementation
- Life Cycle Assessment integration

The proposed sustainable approaches, including the use of PLA biopolymers, vegan leather materials, and agro-waste-derived products, can contribute significantly toward reducing environmental impacts and promoting sustainable textile manufacturing systems. The integration of green technologies and circular resource management strategies is expected to play a major role in the future transformation of textile industries toward environmentally responsible production practices.

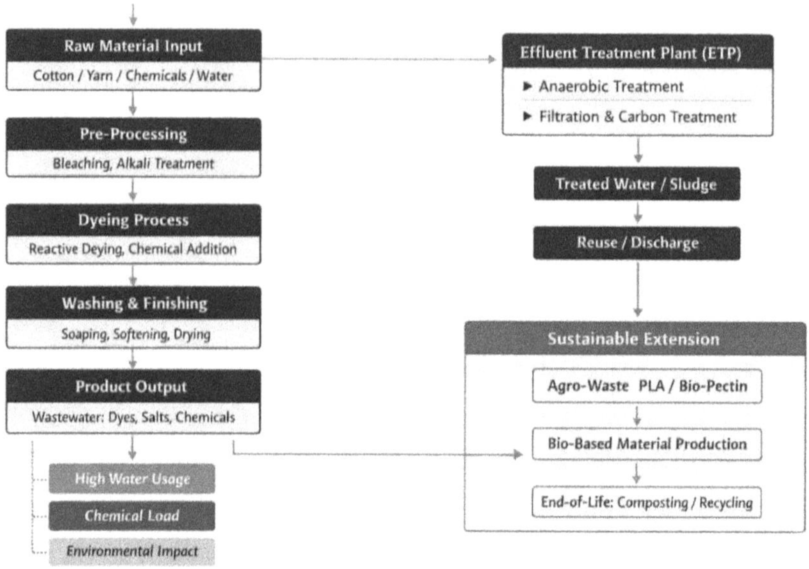

Life Cycle Assessment (LCA) of Textile Processing

REFERENCES

1. Auras, R., Lim, L. T., Selke, S. E. M., & Tsuji, H. (2010). Poly(lactic acid): Synthesis, structures, properties, processing, and applications. John Wiley & Sons.
2. Arora, S., & Mehra, R. (2021). Agricultural waste-based biocomposites for sustainable leather alternatives. Journal of Cleaner Production, 318, 128531.
3. Chen, G. Q. (2010). Plastics completely synthesized by bacteria: Polyhydroxyalkanoates. In Plastics from Bacteria (pp. 17–37). Springer.
4. Cheng, J. J., & Stomp, A. M. (2009). Growing duckweed to recover nutrients from wastewater and for production of fuel ethanol. Clean – Soil, Air, Water, 37(1), 17–26.
5. Ciriminna, R., Fidalgo, A., Delisi, R., Ilharco, L. M., & Pagliaro, M. (2016). Pectin production and global market. Agrochimica, 60(1), 1–12.
6. Cui, W., Cheng, J. J., & Stomp, A. M. (2010). Growing duckweed for biofuel production. Biofuels, 1(5), 691–702.
7. Fishman, M. L. (2011). Pectin and pectin-based biomaterials. In Biopolymers (pp. 181–204). Elsevier.
8. Garlotta, D. (2001). A literature review of poly(lactic acid). Journal of Polymers and the Environment, 9(2), 63–84.
9. Ge, X., Zhang, N., Phillips, G. C., & Xu, J. (2012). Growing Lemna minor in agricultural wastewater and converting the duckweed biomass to ethanol. Bioresource Technology, 124, 485–488.
10. Jamshidian, M., Tehrany, E. A., Imran, M., Jacquot, M., & Desobry, S. (2010). Poly-lactic acid: Production, applications, nanocomposites, and release studies. Comprehensive Reviews in Food Science and Food Safety, 9(5), 552–571.

11. Joseph, M. L., & Nithya, N. (2022). Material flows in the life cycle of leather and vegan leather. Cleaner Engineering and Technology, 7, 100421.

12. Kirchherr, J., Reike, D., & Hekkert, M. (2017). Conceptualizing the circular economy. Resources, Conservation and Recycling, 127, 221–232.

13. Koller, M. (2018). Production of polyhydroxyalkanoate (PHA) biopolyesters from waste streams. Materials, 11(9), 1565.

14. Kral, I., et al. (2023). Sustainable alternatives to conventional leather: Plant-based and biofabricated materials. Sustainability, 15(4), 3215.

15. Kumar, V., Sindhu, R., Binod, P., Pandey, A., & Faraco, V. (2020). Waste biorefineries: Future energy, green products and sustainability. Bioresource Technology, 302, 122841.

16. Lim, L. T., Auras, R., & Rubino, M. (2008). Processing technologies for poly(lactic acid). Progress in Polymer Science, 33(8), 820–852.

17. Lin, C. S. K., et al. (2013). Food waste as a valuable resource for the production of chemicals, materials and fuels. Energy & Environmental Science, 6(2), 426–464.

18. Mohanty, A. K., Misra, M., & Drzal, L. T. (2005). Natural Fibers, Biopolymers and Biocomposites. CRC Press.

19. Paritosh, K., et al. (2017). Food waste to energy: An overview. Renewable and Sustainable Energy Reviews, 73, 1219–1227.

20. Sharma, S., & Chatterjee, S. (2017). Microplastic pollution: A threat to marine ecosystem and human health. Environmental Science and Pollution Research, 24(27), 21530–21547.

21. Sudesh, K., Abe, H., & Doi, Y. (2000). Synthesis, structure and properties of polyhydroxyalkanoates. Progress in Polymer Science, 25(10), 1503–1555.

22. Ubando, A. T., Felix, C. B., & Chen, W. H. (2020). Biorefineries in circular bioeconomy: A comprehensive review. Bioresource Technology, 299, 122585.

23. Verlinden, R. A. J., Hill, D. J., Kenward, M. A., Williams, C. D., & Radecka, I. (2007). Bacterial synthesis of biodegradable polyhydroxyalkanoates. Journal of Applied Microbiology, 102(6), 1437–1449.

24. Xu, J., Cui, W., Cheng, J. J., & Stomp, A. M. (2011). Production of high-starch duckweed and its conversion to bioethanol. Biosystems Engineering, 110(2), 67–72.

25. Zhao, H., Appenroth, K. J., Landesman, L., Salmeán, A. A., & Lam, E. (2012). Duckweed rising at Chengdu. Plant Molecular Biology, 78(6), 627–632.

www.ingramcontent.com/pod-product-compliance
Lightning Source LLC
LaVergne TN
LVHW052114210926
848482LV00025B/622